Electromagnetic Compatibility in Medical Equipment

A Guide for Designers and Installers

William D. Kimmel
Daryl D. Gerke

The Institute of Electrical and Electronics Engineers, Inc.
New York

Interpharm Press, Inc.
Buffalo Grove, IL

Invitation to Authors

Interpharm Press publishes books focused upon applied technology and regulatory affairs impacting Healthcare Manufacturers worldwide. If you are considering writing or contributing to a book applicable to the pharmaceutical, biotechnology, medical device, diagnostic, cosmetic, or veterinary medicine manufacturing industries, please contact our Director of Publications.

Social Responsibility Programs

Reforestation

Interpharm Press is concerned about the impact of the worldwide loss of trees upon both the environment and the availability of new drug sources. Therefore, Interpharm supports global reforestation and commits to replant trees sufficient to replace those used to meet the paper needs to print its books.

Pharmakos-2000

Through its Pharmakos-2000 program, Interpharm Press fosters the teaching of pharmaceutical technology. Under this program, complimentary copies of selected Interpharm titles are regularly sent to every College and School of Pharmacy worldwide. It is hoped that these books will be useful references to faculty and students in advancing the practice of pharmaceutical technology.

Library of Congress Cataloging-in-Publication Data

Kimmel, William D.

Electromagnetic compatibility in medical equipment : a guide for designers and installers / William D. Kimmel, Daryl D. Gerke

p. cm.

Includes bibliographic references and index.

ISBN 0-935184-80-5

ISBN 0-7803-1160-4

1. Electromagnetic compatibility. 2. Medical electronics. 3. Electromagnetic interference. I. Gerke, Daryl D. II. Title.

R857.E54K54 1995

681′.761—dc20 95-25000 CIP

10 9 8 7 6 5 4 3 2 1

This is a joint publication between IEEE Press and Interpharm Press, Inc.

Interpharm ISBN: 0-935184-80-5

IEEE ISBN: 0-7803-1160-4

IEEE Order Number: PC5633

Interpharm Press, Inc.
1358 Busch Parkway
Buffalo Grove, IL 60089, USA
Phone: + 1 + 708 + 459-8480
Fax: + 1 + 708 + 459-6644

IEEE Press
The Institute of Electrical and Electronics Engineers, Inc.
445 Hoes Lane, P. O. Box 1331
Piscataway, NJ 08855-1331, USA
Phone: +1 + 908 + 562-3967
Fax: +1 + 908 + 562-1746

Contents

This book is based on a series of articles that appeared in *Medical Device and Diagnostic Industry* during 1992–1995.

Preface

This book was written in recognition that there is no up-to-date EMC design guide for the medical device designers. We have written this material for the express purpose of providing the designer and installer with a nuts and bolts guide on EMC design and installation. We have minimized the theory and have concentrated on providing rules of thumb on the correct way to ground, shield, and so on, without having to relearn Maxwell's equations.

Our rules of thumb and our examples are based on experiences gained in our consulting practice–being on the firing line. In most cases the designer will be able to make coherent decisions based on the rules provided herein. We recognize that the designer rarely has full control over the design, and often has to make do with a given set of constraints (besides the obvious ones relating to medical devices): Circuit technology has been specified, existing circuit boards must be used, plastic enclosures are specified, and so on. The goal for the designer then becomes one of choosing the best path and seeing what happens.

Where the reader wants more information, there a number of good references available, and these are cited in the Appendix A.

ACKNOWLEDGMENTS

The materials are an outgrowth of a series of articles written for *Medical Device & Diagnostic Industry* (MD&DI). Our thanks to John Bethune and staff for the opportunity to hone our writing skills and for keeping us honest. Without them, this book would not have come into being.

Our thanks to our friends in the EMC business, especially Dr. Tom Chesworth (publisher of *Electromagnetic News Report*) and Bill Ritenour for their insights. Whenever we have a question, they are always willing to share their time and expertise. And a special thanks to Gene Panger for helping us with the chapter on regulations.

Finally, thanks to our wives, Sharon Kimmel and Mary Lou Gerke. They were truly "author widows" during the months in which this manuscript was in process.

1

INTRODUCTION

Modern medical devices are packed with electronics, ranging from sensitive analog amplifiers to sophisticated microprocessors. Unfortunately, these same devices can be adversely affected by EMI (Electromagnetic Interference) problems. Furthermore, these EMI problems can be compounded due to critical missions (human life may depend on proper operation) and harsh EMI environments (ranging from the operating room to emergency vehicles to patient homes). Designers and users of modern medical devices need to know how to identify, prevent, and fix these EMI problems.

Common EMI threats can cause upsets and even damage to sensitive medical devices. In the following chapters we will discuss the physics of EMI, then we will start from the inside of electronics and work to the outside. We will discuss the problems and give you some common cases and effective remedial techniques at three levels–circuit level, box level, and interconnect levels.

The leakage current limit is the single most important EMI–related concern in medical device design because it is almost impossible to adequately filter signal and power lines given the current limit. This major design challenge essentially mandates that EMI issues be addressed from the very start of a design project. Designers need also be aware that medical devices will soon be subject to new EMI regulations in Europe, and that similar requirements may be implemented in the United States. As currently proposed, these requirements reflect the real operating environment, providing helpful guidance for designers. As is true of all design issues, however, the sooner all of these EMI issues are addressed, the less expensive and more successful the final design will be.

WHAT ARE THE THREATS?

EMI has two faces–emissions and immunity (also known as susceptibility). Emissions will come from your equipment, possibly to interfere with the operation of nearby electronic equipment. On the other hand, external electrical energy may adversely affect your equipment. Thus, your equipment may be the source of interference or the recipient. These threats may be propagated in two ways: radiated and conducted, or a combination of the two.

If you are only concerned about design regulatory issues (see Chapter 2), the requirements are easy to identify–just read the specification. These requirements are not arbitrary, but represent hundreds of "engineer years" of research, analysis, and committee work to arrive at these levels. As a result, the recommended levels represent real world threats.

But suppose you have already met your applicable standards, and you are still getting EMI reports from the field. Remember, these standards are only good guides and cannot possibly cope with every condition in the field. Maybe your equipment is placed very close to an ESU (electrosurgical unit), or maybe you have some high level transmitters nearby. You will need to identify the actual threat before you take corrective action.

Let us take a look at the principal threats to your equipment: The sources may be transient in nature or may be nearly continuous. The three common EMI threats are RFI (radio frequency interference), ESD (electrostatic discharge), and power disturbances. Each are common sources of problems to electronic equipment, although they have widely different signatures and typically impact different circuits.

Emissions

Emissions are a measure of electrical energy emitted from your equipment. Some medical devices generate high electrical energy as a part of its function, and this, of course, is a concern. But most electronic equipment generates so little energy that it is a potential threat only to nearby radio equipment. Accordingly, emissions control is primarily a regulatory issue: The emissions may not adversely affect nearby medical equipment, but the general regulatory limits will still apply.

Radio Frequency Interference

As the name implies, RFI deals with threats in the RF, or radio frequency, range. Traditionally, this begins at about 10 kHz on the low end, and usually extends to 500 or 1000 MHz for commercial applications, and to 40 GHz (radar frequencies) for military or aircraft applications. RFI is usually continuous (long relative to circuit response time) rather than transient.

Electrostatic Discharge

Electrostatic discharge follows a gradual charge buildup. Actual charge buildup will occur over a period of time, generally seconds or more, and usually poses no threat to electronics. When discharge occurs, it takes only a few nanoseconds, and this is what causes the ESD problem. Actual discharge can be from a human body to or near the equipment in question, or it may be accumulated by the equipment (as in the case where the equipment is portable or on rollers) and discharged from there.

Power Quality

Power quality, or lack thereof, is simply a deviation from the 50 or 60 Hz ideal sine wave power we have come to hope for (or demand). Power disturbances can have many sources and take many forms, both continuous and transient. Power disturbances are often generated locally by nearby noisy equipment, and, thus, cannot be controlled by the power company. Power disturbances are tough to eliminate completely, so it is usually necessary to cope with them. These disturbances can have a variety of signatures, depending on the nature of the source, but only a few types of disturbances are of significance in electronics.

Self-Compatibility

Self-compatibility deals with threats internal to your medical electronics equipment. Simply, your equipment is prevented from satisfactory operation by an electrical interference generator in your equipment. Two common cases are when you are dealing with very sensitive equipment (sensitive to interference even from ordinary digital electronics) and when you have a high energy generator that is powerful enough to interfere with nearby equipment, including the electronics within. This situation often surfaces when you are integrating purchased modules into your system.

SOURCES, PATHS, AND RECEPTORS

When dealing with any EMI problem, it is important to divide the problem into three categories–a source, a receptor, and a path coupling the source to the receptor. All three must exist for there to be a problem and, therefore, eliminating any one of these will eliminate the problem. But in any specific case, some solutions are more practical than others. For example, if your equipment is upset by a nearby radio transmitter, you usually cannot eliminate the source–but you can use shields or filters to block the path, or you can harden the receptor at the circuit level. So, you attack the problem where you have some control over the outcome. Equipment designers, of course, will concentrate on hardening the circuits at the receptor. Let us look at the common sources, paths, and receptors for each of the threats in turn.

Sources

RFI Sources

What are some typical RF sources? In today's modern society we are literally awash in RF energy from a wide range of sources. These include ***natural*** and ***man-made*** sources, and ***intentional*** and ***unintentional radiators***. Table 1-1 shows some typical high level RF sources we might find in the medical environment.

The electric field strength of an RF source drops off inversely with distance from the source. Thus, a low power walkie-talkie located a meter away poses a much greater threat than a high powered commercial broadcaster a kilometer away. Even worse–this handheld source is mobile, making it difficult to trace.

> For example, a maintenance person keys a handheld radio in the hallway and upsets a sensitive piece of equipment, and then walks away unaware of the problem he or she just caused. Do not overlook this threat–it may explain some mysterious field failures. With the proliferation of handheld radios and cellular phones, we always put these sources on our suspect list.

Mobile radio transmitters are also a threat, particularly near emergency vehicles. These transmitters have higher power than the handheld types and they often pose a threat at distances of 10 meters or more. Any equipment in or near the emergency wing of a hospital will be exposed to these threats on a regular basis. Any medical devices mounted in or carried into an emergency vehicle are exposed to even higher levels of RF energy and must be protected.

Many common medical devices use RF energy and can pose a threat to nearby electronic equipment. These include diathermy units, MRI systems, lasers, and ESUs. The ESU used in an operating room is particularly nasty, as it "sprays" RF all over the place, upsetting even robust electronics, not to mention sensitive monitoring equipment.

Table 1-1. High Level RF Sources

Category	Name	Frequency Range
Medical devices (unintentional radiators)	Diathermy	27 MHz–500 MHz
	ESU	30 kHz–100 MHz
	MRI	60 MHz
	Lasers	27 MHz (varies)
Radio transmitters (intentional radiators)	Television	54 MHz–800 MHz
	Radar	1 GHz–40 GHz
	AM radio	550 kHz–1.6 MHz
	FM radio	88 MHz–108 MHz
	Land mobile*	30–50 MHz
		150–170 MHz
		450–500 MHz
	Cellular phone	900 MHz
Other	Arc welders	2–20 MHz
	RF heaters	13.5, 27, or 40 MHz typical

*Land mobile includes police, fire, ambulance, pagers, and walkie-talkies.

Other potential RF threats include theft detectors, RF welders, and RF heat-sealing equipment. While not commonly found in the medical environment (although arc welders may be involved in hospital construction projects), they can cause problems to patients equipped with portable electronics, and you should be alert for these unexpected threats.

The measure for RF problems is the "electric field intensity," given in volts/meter. This can be measured with a field strength meter, and, in simple cases, can also be predicted. For today's electronics failures typically occur in the 1–10 volts/meter range, although we have seen failures on some sensitive systems in the 0.1 volt/meter range. Unfortunately, nearby radio transmitters can cause levels in the 1–100 volts/meter range, depending on the power level and the distance from the transmitter. This is why most medical RF regulations now specify test levels in the 3–10 volts/meter range. But even this is not enough for equipment used in emergency vehicles carrying radio transmitters, which may experience field strengths as high as 200 volts/meter.

ESD Sources

ESD requires charge buildup before discharge can occur. Actual charge buildup originates by rubbing two materials together (at least one of which is a dielectric), resulting in an accumulation of positive charge on one material and a negative charge on the other. A useful table, called the Triboelectric Series (Appendix B), places common materials on a relative scale, from the more electropositive to more electronegative.

The farther apart on the scale, the more readily charge will accumulate. But do not place undying faith on Table 1-2. Charge will accumulate when rubbing two similar or even identical materials together, as evidenced by substantial charge buildup when unrolling homogeneous tape or paper.

As voltage is building up, it is also bleeding off, either through the material or through the air. Bleed rate is heavily influenced by humidity, as well as the resistance of the media in question. Even poorly conductive materials (volume resistivities of 10^6–10^{11} Ω-cm) are usually able to bleed off charge before it builds up to excessive levels. In moist climates (relative humidity 50 percent) electrostatic buildup is relatively low and detectable discharges are unusual. In dry climates (relative humidity 10 percent), electrostatic charge can quickly build up to levels sufficient to elicit vicious sparks.

When the electric field intensity becomes high enough, breakdown occurs, which results in an abrupt rush of current accompanied by a collapsing electric field. Discharge occurs when the field strength reaches a high enough level to allow arcing, either through air or other media. Generally, this occurs when a charged body is brought into close proximity to ground. A 2 kV discharge is about the threshold of feeling but is enough to destroy many electronic components. Actual voltages can build up to 15 kV (more in certain circumstances), which is quite a vicious spark.

The source of ESD is always due to rubbing two materials together. Even air flowing over certain materials is capable of transferring enough charge to destroy unprotected electronic components, but usually cannot build up enough charge to affect assembled equipment.

Most ESD problems arise as a result of discharge from a human. Another common problem is discharge from furniture, notably from a table or a device on wheels. Less common is the moving member problem, such as from belt-driven devices or film rollers. Common situations include a human walking across the floor, scuffing one's feet on tile or, worse yet, carpeting. The human picks up charge, leaving an oppositely charged spot in his or her trail. The human body can be modeled by a capacitance of about 150 pF and a resistor of 330 ohms. The capacitance is primarily that between the person and the floor and is dependent on the properties of the shoe sole. Women, commonly wearing thin-soled flats, have a higher capacitance than men, and are a bigger ESD threat (we always thought women were electrifying!).

Rubber rollers, such as those used on a cart, are excellent sources of charge pickup. A common sight in a hospital is the patient walking down the hall, trailed by an infusion pump, mounted on (you guessed it!) rubber rollers.

Moving members and materials in the equipment itself may generate static charge. Examples are paper rolling out of a printer and internal belts associated with motors or sliding members. Tread mills are a copious source of charge buildup.

The actual buildup is not usually harmful to electronics–it is the discharge that does the harm. The discharge from a human body is modeled as in Figure 1-1. The discharge is characterized by a 1 ns rise time to a peak current of about 10 amp. This 1 ns rise time equates to an equivalent frequency of 300 MHz.

Power Disturbance Sources

The source of power disturbances is widely varied. Figure 1-2 shows several characteristic signatures of power disturbances.

Voltage Variations. There are a number of possibilities of voltage variations. Sags and swells result in the brief voltage variations that occur due to temporary load variations nearby. An example would be the starting or stopping of a heavy load, such as an elevator motor. These show up as the brief dimming or brightening of incandescent lamps. Overvoltages and undervoltages are out of tolerance line voltages caused by long-term load conditions, the cause of which is usually outside your facility. A half-cycle voltage dropout may occur due to a crowbar transient device firing. Outages, of course, are a result of the complete loss of power for at least some number of cycles.

These variations are an issue of energy availability. Even in well-controlled environments, sags and outages can occur. Modern switching power supplies can withstand a considerable amount of

Figure 1-1. Waveshape of ESD from a Human Body

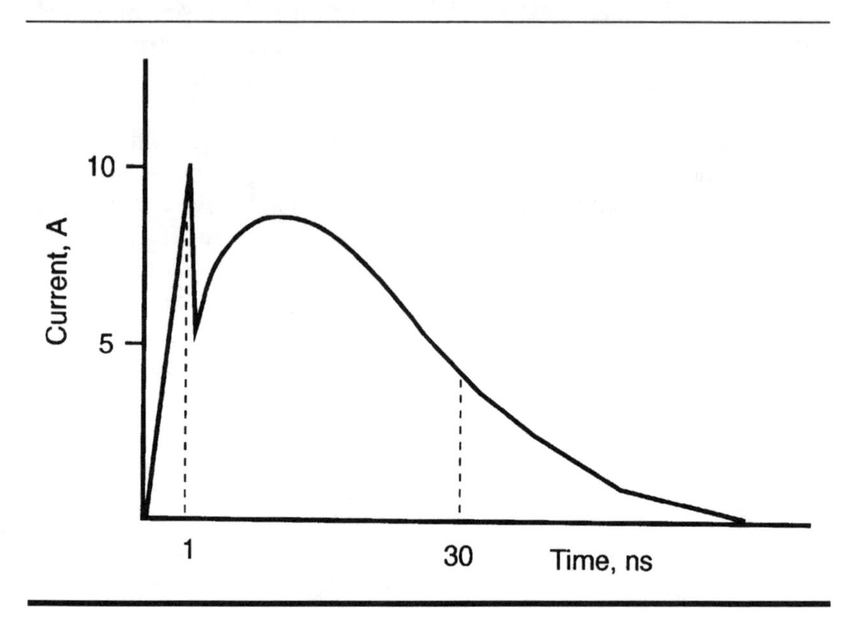

sag or swell, but linear supplies are much less tolerant. Even switching power supplies have their limitations, however, and may need the assistance of an external power conditioner or power source.

Frequency Variations. Power line frequency that is higher or lower than the nominal 50 or 60 Hz is a frequency variation. In the U.S. and in most industrialized countries, the power line frequency is quite stable. In underdeveloped countries and small emergency power sources, the frequency is much less well controlled. In any event frequency variations rarely cause a problem with power supplies, except for ferroresonant supplies–not widely used in electronics, but common in UPS.

Waveform Distortions. An extreme case of nonlinear loading from electronic equipment, commonly in computer equipment, produces waveform distortions. Power supplies, particularly switching supplies, draw current from the middle of the AC cycle, causing flat topping of the voltage wave. This condition causes heating problems with transformers, motors, and wiring, but rarely causes a problem with electronic power supplies.

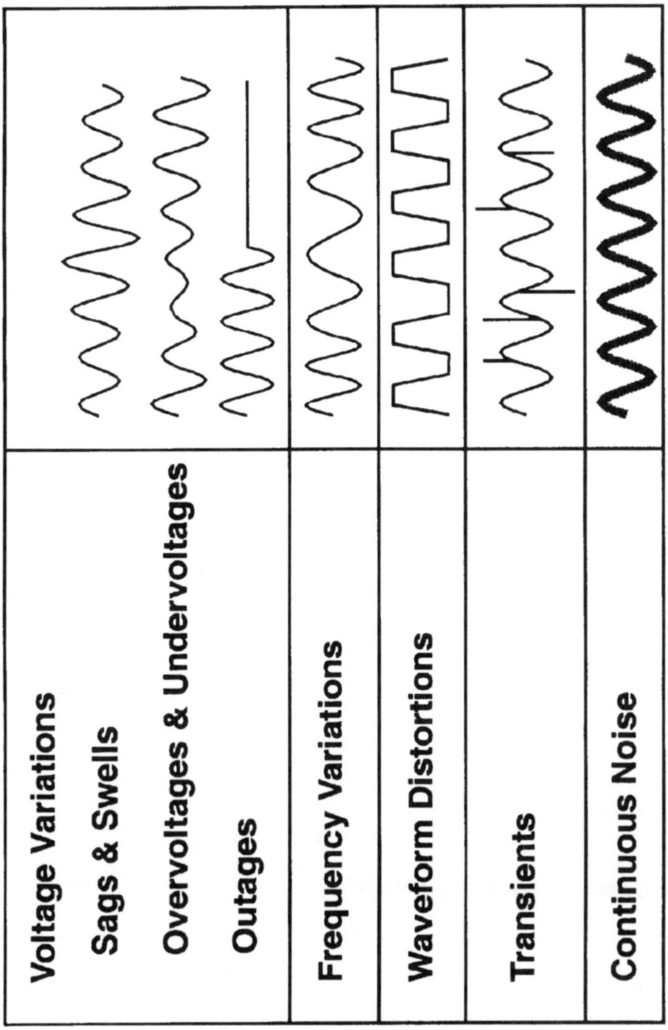

Figure 1-2. Power Disturbance Signatures

Voltage Variations **Sags & Swells** **Overvoltages & Undervoltages** **Outages**	
Frequency Variations	
Waveform Distortions	
Transients	
Continuous Noise	

Transients. Transients are generally in two categories: Lightning surge transients and EFTs (electrical fast transients). The lightning transient is really high energy and may be either from a direct lightning strike or by induction from a nearby strike. The wave on the line may be an overdamped transient or oscillatory, depending on various line conditions. This is a severe requirement, often called a smoke test, as the test criteria may be that the equipment not be left in an unsafe condition (shock or fire hazard). Lesser amplitude transients, of course, must be tolerated.

EFTs usually occur from interrupting current to a nearby inductive load, resulting in a rapidly rising voltage spike, usually following by damped high frequency ringing. This high frequency burst can be devastating to power supplies, which are designed to filter low frequency ripple, but usually not the high frequencies described in this requirement. This condition occurs quite often and equipment should be designed to withstand this transient.

Continuous Noise. Radio frequency interference that may be generated by nearby radio equipment, especially an AM radio, often causes continuous noise. Radio frequency interference, as described above, may enter your equipment via the power line. On an oscilloscope, RFI appears as a fuzzy sine wave, but is actually high frequency energy imposed on the power line voltage.

Generally, electronic equipment should be able to withstand modest amounts of interference. Figure 1-3, adapted from FIPS Pub 94 and the IEEE "Emerald Book," shows a recommended interference tolerance vs. duration of the disturbance. Long-term disturbances are simply operating voltage tolerance. As the disturbance duration becomes shorter, the recommended interference amplitude tolerance increases, finally reaching several thousand volts for very brief transients.

Interference Paths

Interference paths may be conducted, radiated, or, most often, a combination of the two. Furthermore, these conducted paths can involve any conductor–power lines, signal lines, and ground paths, not to mention the fortuitous conducting paths in any facility: building steel, conduit, plumbing, and plenums.

RFI Paths

The coupling mode depends on frequency and wavelength. Low frequencies travel easily along conducted paths, but do not radiate

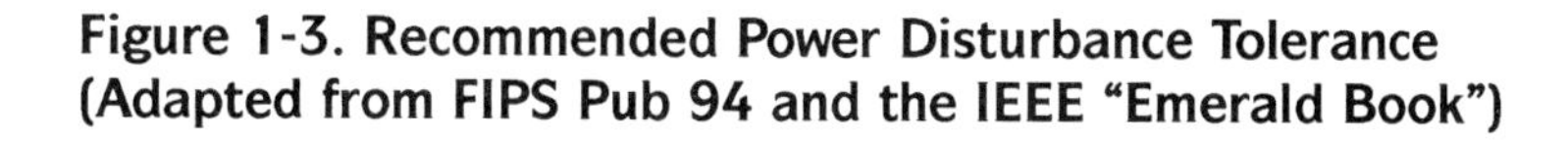

Figure 1-3. Recommended Power Disturbance Tolerance (Adapted from FIPS Pub 94 and the IEEE "Emerald Book")

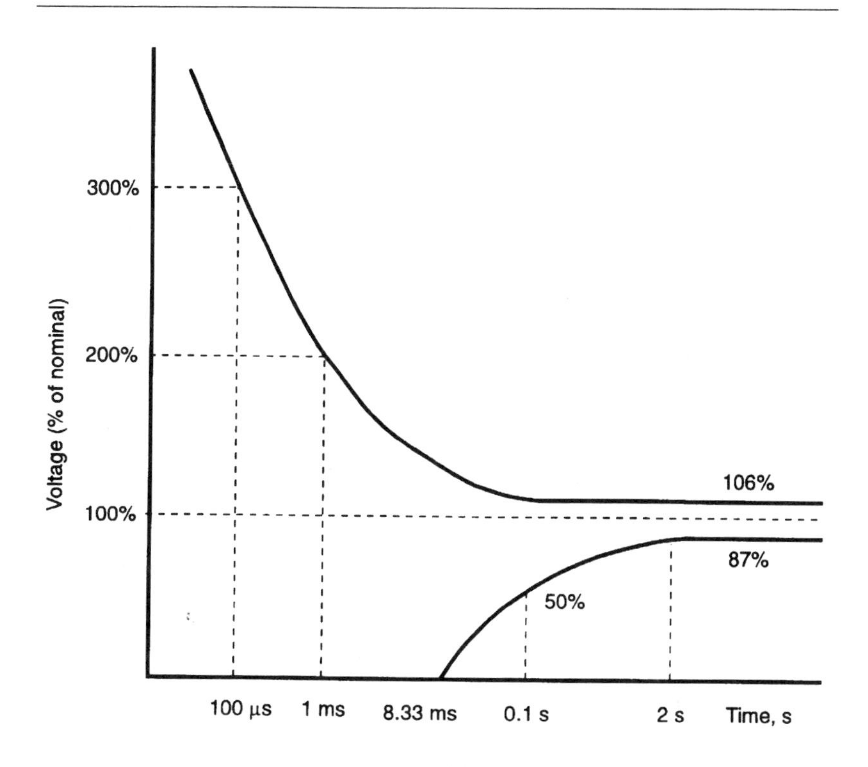

efficiently. High frequencies radiate efficiently, but are impeded by inductance in wires.

What is low and high frequency? A commonly applied metric is to assume that when a cable (or wire) is longer than 1/20 of a wavelength, the cable or wire becomes a good antenna, so the radiated path will predominate. Below 1/20 of a wavelength conducted effects will tend to dominate. Table 1-3 shows the relationship between frequency and wavelength. As you can see, effective radiated coupling from an AM radio transmitter at 1 MHz needs long cables, but above 30 MHz even a few cm may be enough to act as an unwanted antenna.

We can extend this criteria a bit further to cabinets and enclosures, too. Any openings greater than 1/20 of a wavelength can act

Table 1-3. Relationship between Frequency and Wavelength

Frequency	Wavelength*	1/20 of a Wavelength
1 MHz (AM broadcast)	300 m	15 m
30 MHz (land mobile)	10 m	50 cm
100 MHz (FM broadcast)	3 m	15 cm
150 MHz (land mobile)	2 m	10 cm
450 MHz (land mobile)	75 cm	4 cm

*Wavelength can be computed from: λ (meter) = 300/f (MHz)

as "slot antennas," and can actually couple reradiated energy from outside the box to the internal electronics. As you can see, at 450 MHz, a 4-centimeter slot (less than 2 inches) can pose problems.

ESD Paths

As will be shown in chapter 3, the 1 ns rise time of an ESD wave equates to an equivalent frequency of 300 MHz, and ESD will follow much the same path as RFI at that frequency. High-frequency energy does not travel well down a wire, due to inductance in the wire, but it does couple very well.

ESD effects may be due to direct or indirect discharge. For direct discharge the most devastating ESD path is direct discharge to a signal wire. Direct discharge to a ground wire is almost as bad. A damaging ESD event needs to be fairly close, otherwise lead inductance will block the devastating peak currents.

An indirect discharge path is primarily via a magnetic field. The fast rise of the discharge path creates an intense magnetic field that can couple into nearby circuit loops. A small circuit loop can easily generate enough to upset most logic. This is a very common condition for discharge near a plastic enclosure.

Another indirect discharge path is to a metal plate, in which case the principal effect is the electric field. A common path is discharge to an enclosure door, which occurs when the door is poorly grounded to the enclosure. We have observed effects from such discharges at a distance of 5 meters.

Power Disturbance Paths

Unlike ESD and RFI, which usually have multiple paths, power disturbances are always conducted through the power line. The disturbance can be line to line, or it can be line or neutral to ground. In fact, there is always copious amounts of ground noise within a facility, even if the power source is very clean, ground interference may still cause problems.

Receptors

Receptors of RFI

Circuits are the ultimate receptors of RFI. But which are the most vulnerable? Most electronic devices today contain digital circuits, but many medical devices often also contain sensitive analog circuits. Digital circuits have much higher noise immunity levels, but can also be upset, given a strong enough source.

Analog circuits are much more susceptible to RF than digital circuits. The most sensitive amplifiers are usually at the front end, which, unfortunately, are often connected to patients via cables (these are the unwanted antennas.) Common characteristics of such inputs are low signal levels (10 μV to 10 mV), high input impedance (1 MΩ or more), low bandwidth (under 10 kHz), and extremely low leakage currents (20 microamps). The leakage current is particularly vexing, as it can severely limit filtering and decoupling.

Analog circuits are subject to "audio rectification." Analog amplifiers detect and demodulate RF. If these demodulated signals fall into the expected signal frequency range, they will be processed as "real" signals. This explains why high frequency sources (such as a 150 MHz transmitter) cause problems with a low frequency receptor (such as an ECG channel). Once this demodulation occurs, no amount of filtering can remove the offending signals. Incidentally, since rectification occurs due to nonlinearities, bipolar devices are more prone to these failures.

We have also seen RFI problems with power circuits, such as regulators. These devices use feedback, and even a small amount of RF in the feedback loop can cause big problems in the power output.

Receptors of ESD

ESD will cause both momentary upset or permanent damage to electronic components. Indirect discharge is most often characterized by transient effects, from which the circuit quickly recovers. Permanent damage often arises from direct discharge, which has typically much greater amplitudes. Let us take a closer look at each.

Permanent damage is a widely recognized problem with modern insulated gate field effect devices, as even a few hundred volts is sufficient to cause these devices to fail catastrophically. Another possibility is for the ESD to drive the circuit into latch-up, sometimes destroying the part (the chip overheats) or sometimes simply biasing the chip into a nonfunctional state. Other devices often do not fail catastrophically and join the ranks of the "walking wounded."

> For example, an overvoltage at a PN junction will punch through and damage the junction without destroying it. This results in a shift in the performance parameters of the device, notably leakage current, which passes through the crater, causing local hot spots. Additional ESD events punch additional holes in the junction, increasing the leakage current. Eventually, the junction fails due to overheating.

Internally, some ICs have built-in protection for overvoltages, which is better than nothing, but are often ineffective. Even if they do their job, remember that the ESD current is being shunted down some awfully fine metallizations and lead bonds. If the metallization opens, the component fails, of course, and leads to an obvious malfunction. But the metallization may simply be eroded, reducing the current-carrying capability, ultimately to fail due to other apparently unrelated causes. Overvoltage conditions can occur between any set of pins on the chip, including V_{CC} and ground, but the most common is transients to signal lines that go appreciably beyond the V_{CC}/ground limits.

Transient effects to active electronics are much more prevalent. Digital circuits are very vulnerable to ESD hits. Modern circuits are quite capable of responding to the narrow transient, amplifying it, and passing it on, ultimately latching it in a storage register or memory cell. In such cases external intervention is needed to restore the register or memory cell to a legitimate value. Analog circuits are fairly immune to ESD signals (but not for damage), although we have also seen upset in high impedance feedback circuits.

If latch-up occurs, but does not destroy the chip, it must be powered down before normal operation can be resumed (still, that is better than replacing the chip). One of the unusual effects of latch-up on modern microprocessor chips is that a portion of the chip may latch without affecting the rest of the chip. You may lose the function of a single channel, for example, with today's ESD and another channel tomorrow.

Passive components are also subject to ESD. A sufficiently high ESD voltage impressed across the leads of a thin film resistor will often result in a arcing path being established around the resistor. The thin varnish insulation in magnet wire is incapable of withstanding ESD voltages. Accordingly, ESD applied to an inductor or transformer usually arcs across the windings, perhaps creating a shorted turn in the process.

Receptors of Power Disturbances

Power disturbances may affect the power supply, or may pass through to attack the electronics within, causing temporary upset or, in severe cases, component destruction. For most power disturbances the issue is one of energy storage and regulation. If the power supply cannot feasibly be designed to handle these conditions, then external conditioning will be required.

In the case of a high voltage transient caused by lightning, the principal risk is an overstress to electronic equipment, especially the power supply. This is simply a case of too much voltage and current for the supply to handle. For severe transients it may be sufficient that the equipment is not left in an unsafe state (i.e., on fire or with dangerous voltages). For lesser transients it is desirable that the equipment survive and operate through the transient without adverse effect.

The most common power disturbance problem is with EFTs, where a high frequency burst enters via the power line to attack internal electronics. The recipient is either the power supply or the electronics within. In the power supply the recipient is often the voltage regulator.

Alternately, the transient will pass right through the power supply to attack the circuits within, particularly resets. Power supplies are designed to filter out low frequency ripple with low frequency electrolytic capacitors, and these capacitors will not handle the high frequency burst. While a power line filter will catch much of the transient, the disturbance amplitudes are high enough that enough sneaks through to upset internal electronics.

Transient effects to internal electronics are also common. Digital circuits are very vulnerable to EFTs, capturing the transient, ultimately latching it in a storage register somewhere. Often the reset circuits are involved. In such cases external intervention may be needed to restore the system integrity. Analog circuits are vulnerable to RFI that may enter through the power supply, but are fairly immune to power transients, although we have also seen upset in high impedance feedback circuits.

In extreme cases an overvoltage will stress the component, causing damage. Occasionally, latch-up will occur, sometimes destroying the chip. The symptoms are similar to ESD, and it is often difficult to determine what is really causing the problem. If latch-up occurs, but does not destroy the chip, it must be powered down before normal operation can be resumed. Latch-up may affect only a portion of the chip while the rest remains fully functional.

SUMMARY

EMI poses five major threats to medical electronic equipment: Emissions, RFI, ESD, power disturbances, and self-compatibility.

1. Emissions are primarily a regulatory issue.
2. RFI primarily impacts analog circuits.
3. ESD primarily impacts digital circuits.
4. Power disturbance problems are primarily due to spikes and sags.
5. Self-compatibility problems are primarily due to high energy jamming all electronics, and digital electronics affecting sensitive analog electronics.

All interference problems have a source of interference, a path of interference, and a receptor of interference.

- Common sources include diathermy, ESU, MRI and lasers, walkie talkies, cellular phones, welders, and RF heaters.
- Paths may be radiated, conducted, or a combination.
- Conducted paths dominate below 30 MHz; radiated paths dominate above 30 MHz.
- Receptors of EMI include digital circuits, vulnerable primarily to transients, and analog circuits, vulnerable primarily to continuous wave interference.

2

EMI STANDARDS FOR MEDICAL DEVICES

There is a flurry of activity from standards committees to formulate and adopt EMI performance standards for medical electronics. These standards are intended to limit emissions from your equipment and to establish the minimum levels of interference that your equipment must tolerate.

As medical electronics has proliferated, we have found electronics being operated in increasingly hostile environments in the hospital, clinic, and residence. As would be expected, with increasing equipment usage, there are increasing complaints of equipment anomalies and actual failures. There is no doubt that EMI can and does cause problems. Sadly, there is also no doubt that many claims of interference are without basis.

Nevertheless, standards are being adopted increasingly, and we need to be cognizant of the current regulations along with where the world is going in the future. Even more importantly, as designers, how do these regulations impact our design?

This is a book on design, not regulations, so we will not attempt to completely cover the regulations in existence, much less those that are in process. But we will give you some insights as to what the regulations are, where they come from and where they are going.

The IEC EMI requirements serve as the basis for all the current and pending EMI requirements. The primary requirements include emissions, radio frequency interference, electrostatic discharge, and various forms of power and signal line interference.

The actual requirements are dependent on the equipment category, whether the device is invasive, or patient connected or neither. Failure criteria has not been well established, but the direction is to define failure as loss of *clinical utility,* as will ultimately need to be decided by clinicians. Brief upsets may be tolerable, providing equipment self recovers without losing active parameter (i.e., resetting to default parameters would usually not be acceptable). Where analog devices are involved, an operating tolerance may need to be specified.

EMISSIONS VS. IMMUNITY

Both emissions and immunity are of interest to the equipment designer, but for different reasons. Emissions standards limit the emanations from your equipment. Some very sensitive equipment is very vulnerable and may well benefit by emission controls on adjacent equipment, but the emission limits were originally intended for protecting licensed radio communications from interference.

Some medical equipment generates high energy radio frequencies (RF) as part of its function, but most electronic equipment emanates too little energy to affect digital electronics or even most analog electronics. But radio devices are very sensitive, responding even to very low levels of energy, and will receive radio energy from most modern electronics, such as a personal computer. Even so, field strengths fall off reciprocally with distance, and typical test distances range from 3 to 30 meters.

The specified emission level is typically not too demanding, as the primary recipient of radio interference is broadcast radio and TV. These generally operate on considerable field strength, and it takes a lot of energy to cause actual interference. True, there are much more sensitive radios around, and these are much more subject to interference, but these devices are usually isolated from mainstream electronics, and can be treated on an individual basis.

Immunity is an issue that concerns external interference that impairs the function of your equipment. Such interference can arise from a variety of sources, but there are a few characteristic signatures that have been categorized and for which standards have been formulated. These cases include RFI, ESD, and several types of power disturbances. Typical levels have been studied exhaustively, and standards have been formulated.

By the way, you may hear the term *susceptibility* instead of *immunity*. These terms really refer to the same thing. Your equipment will be susceptible above a certain level of EMI, and immune below that level. So you are really talking about the same boundary, and the terms are usually used interchangeably. The military has used the term *susceptibility* for many years. In recent years, the term *immunity* has come into common usage, mostly as a result of the European Union (EU) activity.

THE EU AND THE U.S.

The European Union (EU), formerly called the European Community (EC) or the European Economic Community (EEC), is vigorously imposing EMI regulations on everyone who expects to sell any electrical or electronic equipment in Europe. Both emissions and immunity standards have been formulated and are in the process of being enacted.

The U.S., on the other hand, has no united front. The FCC (Federal Communications Commission) regulates emissions to protect communications, but exempts numerous industries (including medical equipment operated by a professional) from test requirements. This is not to say that excessive emissions cannot cause problems attracting the attention of the FCC, but these are relatively infrequency, and can be handled on an individual basis. And when problems do arise, they are usually from intentional radiators that apply RF external to the enclosure itself, not from the parasitic emissions of most electronic equipment.

The FCC does have authority to regulate immunity levels, too, but there has been no firm move to do so. Still, the subject keeps coming up, and we expect the FCC might change this in the not too distant future. Although there has been vigorous opposition from to regulations from US industry, an informal survey taken at the 1994 IEEE EMC Symposium indicates that attitude may be changing.

The FDA (Food and Drug Administration), on the other hand, has become increasingly concerned about immunity to medical electronic equipment, in view of the often critical mission. This has been driven by horror stories of equipment failure due to EMI. FDA reviewers are increasingly turning to a draft reviewers guideline for assurance that a device meets the EMI threats it is presumed to operate around. This reviewers guide, as a result, has become a de facto mandatory standard, and there is little question that it had better be followed.

Standards in the U.S. and abroad are generally formulated by non-government agencies, typically committees formed up of industry and government representatives. Subsequently, governments enact these standards as regulations or law. This results in a potpourri of standards, some of which are identical.

Although the U.S. and the EU are generally the most active in EMC regulations, other countries are following closely behind. Generally, Canada (via Canadian Standards Association, or CSA) and Mexico follow U.S. standards with little modification, but the rest of the world generally follows the EU requirements.

THE EU

The EU has been working to harmonize standards of member nations (currently numbering 18). According to agreement, all member countries must enact these standards as law and also must accept work done by other member nations (or other countries outside the EU).

EU Directives

As part of this process, the EU has adopted a number of standards for immunity and emissions. To the greatest extent possible, the EU adopts existing standards, as formulated by the IEC. The IEC has representatives from most of the industrialized countries (the United States is represented by ANSI). Nevertheless, such standards are not always available in a timely fashion, so the EU departs from this practice quite often. Let us take a look at what is happening:

The EU is in the process of formulating a set of regulatory standards for everything (even if the need is not clear to us). In the future there will be standards covering many specific types of electronic equipment. At present, the generic standards apply to most equipment, and these assemble a number of basic standards, as described below. Ultimately, product standards will be developed, which will supersede the generic standards. If your product is not covered by a product standard, you will need to default to the generic standards. At the present time there are only a few product specific standards, including medical devices. In the future specific standards will be defined, as time permits, adapting the generic standards to suit. The standards are not necessarily more stringent, but rather, are intended to provide a better standard for unique operating conditions.

The EU has defined three directives: the Medical Device Directive, the Active Implantable Device Directive, and In Vitro Device Directive. There are also categories within each directive, depending on the application.

You will see references to such standards starting with the prefix EN, IEC, or CISPR (International Special Committee on Radio Interference, CISPR being the initials in French). While they are often used interchangeably, it is important to understand the distinction. IEC (and subcommittee CISPR) is an international organization with representatives outside the EU, including the U.S. It has no legal status in the EU or anywhere else.

The EN prefix indicates the standard is recognized with the EU. CENELEC, the official EU committee for electrical and electronic standards, adopts existing standards wherever possible, notably IEC or CISPR standards, but can and does alter them as needed. There are four types of standards cited in the EMC directive: Basic, generic, product family, and product specific. Generally, the EU is working toward product specific standards, but at this time, most of these are not in existence, so the requirements default to a generic standard. Medical devices have their own product specific standard for many of the categories. Let us start with the basic standards and work our way up.

Basic Standards

The basic standards are those prepared by the IEC (and subcommittee CISPR). They include both emissions and immunity, and form the basis for all the standards in the EU, and to a lesser extent, the U.S. Once you understand these, you have the essentials for the generic and product specific standards. These standards are as follows.

IEC 1000-4-2, Electrostatic Discharge

IEC 1000-4-2 covers the immunity of assembled equipment from adverse ESD. This requirement, dated 1995, is identical to and replaces the 1991 version of IEC 801-2, which, in turn, replaced the 1984 version of IEC 801-2. The principal difference is that the 1984 version specifies a 5 ns rise time, and the new version specifies a 1 ns rise time, which is a much more devastating test.

Electrostatic discharge testing is done with an instrument (often called an ESD gun), and can be performed in one of two methods: air discharge or contact discharge. Air discharge means that the gun is charged and then moved into proximity of the equipment where the discharge occurs. This has proven to be an unrepeatable test, and has fallen out of favor. IEC 1000-4-2 emphasizes the contact test. Contact discharge holds the tip of the gun in contact with the equipment and then an discharge internal to the gun occurs. This test is much more repeatable, as it is independent of humidity, rate of approach, angle, and other variables, and is now the preferred method. Obviously, this requires accessibility to bare metal members. A table of test levels is provided in IEC 1000-4-2, and is repeated as table 2-1 for reference. The relationship is nonlinear. This is due to corona effects that become significant above about 4 kV.

Level three is expected to be the norm for European applications. We believe this is inadequate for use in the U.S., where level four is more appropriate. It appears that the U.S. is the only heavily industrialized country where low humidity conditions are widespread. Humidity levels consistently in the 50 percent range would essentially eliminate ESD as a problem.

The actual waveshape of the discharge is a fast rise (1 ns) to a peak level, followed by a slow decay, and is intended to best model discharge from a human. The key elements of the waveshape are its rise time (1 ns) and its peak current (up to 100 a for a 15 kV air discharge). As will be shown in chapter 3, this is a difficult test. Principal recipients are data cables and operator contacts, each of which can serve as a conduit to internal electronics. Indirect tests to a ground plane at the perimeter of the equipment is prescribed, most notably for plastic enclosures where direct discharge is not possible.

Often ESD susceptibility occurs in "windows": A device may operate normally at 4 kV, fail at 6 kV and 8 kV, then operate normally at 10 kV to 15kV. Therefore, it is appropriate to test at intervals from low voltage to high voltage.

A failure definition must be applied, too. A transient effect (say a few seconds) is generally tolerable (a brief flare on a CRT screen would be an example of this). The next level up is a soft failure, which can be corrected by operator intervention, followed by a soft failure that results in loss of data. Computing devices usually do not self-recover from ESD effects, but often programming around will help. Finally, the hard failure requires equipment repair.

In the U.S., ESD is cited by ANSI C63.16, which is in draft at this time.

IEC 1000-4-3, Radio Frequency Interference

IEC 1000-4-3 covers the immunity of electronic equipment from external RFI. This requirement, dated 1995, replaces the 1992 version of IEC 801-3, which, in turn, replaced the 1984 version of IEC 801-3. The 1984 version of IEC 801-3 does not specify a modulation

Table 2-1. Test Levels in IEC 1000-4-2

Level	Air discharge	Contact discharge
1	2 kV	2 kV
2	4 kV	4 kV
3	8 kV	6 kV
4	15 kV	8 kV

level, which we consider a definite deficiency. Our tests indicate that modulated waves have significantly more adverse impact on electronics than unmodulated waves. Nonlinearities demodulate the wave, providing low frequency shifts in sensitive analog input circuits. This deficiency is corrected in IEC 1000-4-3, where an 80 percent modulation is specified at frequencies within the bandpass of the equipment. IEC 801-3 (1984) test frequency range was 27 MHz to 500 MHz, which was changed to 26 MHz to 1000 MHz in the 1992 version.

The field strength requirement is given in table 2-2. Most equipment will be tested to 3 V/m, except for the ISM (Industrial, Scientific and Medical) frequencies where the limit is 10 V/m. We feel these levels are sufficient for most clinical applications, nevertheless, field strengths can and do go higher as with a cellular phone or vehicle radio transmitter. The one V/m level is expected to be limited to sensitive equipment that cannot readily be immunized to higher levels.

Three V/m is a level that most digital equipment will withstand without impact, but above this level, digital devices start to be affected unless extra care is used in the design. Sensitive analog devices are much more likely to be affected by RFI, however, and compliance may require significant efforts.

IEC 1000-4, Electrical Fast Transients

IEC 1000-4-4 covers the immunity of electronic equipment from electrical fast transients. This requirement, dated 1995, replaces the 1988 version of IEC 801-4. Electrical fast transients occur when disconnecting a nearby inductive load, inducing an oscillatory burst on the power line and (to a lesser extent) signal lines. The burst is a series of high frequency pulses, each with rise time of 5 ns and a pulse width of 30 ns, which we consider to be roughly equivalent to a 50 MHz burst. Amplitudes are given in Table 2-3.

Table 2-2. Field Strength Requirements for IEC 1000-4-3

Level	Test Field Strength (V/m)
1	1
2	3
3	10
x	special

Table 2-3. Amplitudes of EFTs in IEC 1000-4-4

Level	On Power Supply	On I/O
1	0.5 kV	0.25 kV
2	1 kV	0.5 kV
3	2 kV	1 kV
4	4 kV	2 kV
x	Special	Special

Typically, level 2 will apply. We find this is a reasonable level as such levels are not uncommon in the field. This test is not difficult to meet, providing adequate high frequency filtering is allowed. Unfortunately, patient connected devices limit leakage currents making adequate filtering extremely difficult to accomplish.

IEC 1000-4-5, Surge

IEC 1000-4-5 covers the immunity of electronic equipment from an electrical surge, as would be experienced with a lightning bolt. This requirement, dated 1995, replaces the 1993 version of IEC 801-5. The pulse has a 1.2 μs rise time and a 50 μs duration. The voltage levels are given in Table 2–4.

Level 3, will be most commonly cited, which is a 2 kV open circuit voltage for common mode surges. When discharge occurs, the short circuit current gets up in the kA range, which is a significant threat. Facility lightning requirements (ANSI C62.41 or IEEE STD 587) specify much higher levels, but equipment is not expected to survive those higher levels.

Table 2-4. Test Levels for IEC 1000-4-5

Level	Voltage
1	0.5 kV
2	1 kV
3	2 kV
4	4 kV
x	Special

Future IEC 1000-4-x Requirements

A number of requirements are in draft, and can be expected to be added in the future, These include IEC 1000-4-6 (covering conducted RF immunity), IEC 1000-4-8, 9, and 10, (covering magnetic field immunity), and IEC 1000-4-11 (covering power line quality). These requirements are still in draft. The first to appear will probably be IEC 1000-4-6 which is intended to accounted for the fact that radiated interference at low frequencies is not real easy to test due to long wavelengths. Accordingly conducted tests will probably be run from 9 kHz to 80 MHz.

CISPR 11, Emissions

CISPR 11 limits both radiated and conducted emissions. These limits are primarily for the purpose of minimizing interference to radio reception, and are scaled for that purpose. The specified levels have no real relevance to other electronic equipment, and no attempt should be made to apply this requirement to electronics other than to protect radio-receiving equipment.

The earlier version of CISPR 11, dated 1975, has been superseded by the 1990 version. It defines two groups and two classes. Group 1 is for equipment that uses RF energy internally only, and has emission limits the same as for CISPR 22 (which is for information technology equipment). Group 2 is for equipment that uses RF energy externally, and is characterized by quite high levels of energy. Specific frequency ranges have been set aside for equipment that depends on radio energy for its function. In these frequency ranges allowed radiation may be unlimited.

Class B equipment is intended for residential use and for those cases where equipment may be sharing power with a residence. Class A is for all other areas. At present, the general interpretation is that medical equipment should always be in Class B, but that interpretation is being debated.

As implied, radiated emission tests are made with an antenna placed either 10 or 30 meters away (depending on the equipment category). The mission is to detect levels of radiated energy, generally from 30 MHz to 1 GHz (frequencies above and below this range are under consideration). Typical test levels are 30 to 37 dBμv/m, depending on the frequency, at a distance of 10 or 30 meters, depending on the equipment category.

Conducted emission tests cover the frequency range of 150 KHz to 30 MHz to levels of 60 to 67 dBμv, depending on the frequency. Conducted emission tests are performed on power line only. Note the conducted limits stop where the radiated limits start. This is in recognition of the fact that low frequency interference is most efficiently propagated by conduction and high frequency interference is most efficiently propagated by radiation. The approximate crossover is 30 MHz, hence the regulations.

IEC 555, Power Factor

This requirement is in two parts, IEC 555-2, Harmonic Content, and IEC 555-3, Voltage Fluctuations. Harmonic content is actually a part of the safety requirements in the EU, but does inevitably impact EMI at the power line. This document limits the amount of power line harmonics that can be generated at the power supply. For low power equipment the requirement is modest; but for high power devices, a significant problem will occur, as the harmonic content is an absolute maximum, not a relative maximum.

Generic Standards

Generic standards are written to cover equipment for which no product specific standards exist.

EN 50081

The generic standard for ISM (Industrial, Scientific, and Medical) and commercial equipment is EN 50081-1 for emissions and EN 50081-2 for immunity. Lacking any other guidance, you would select these requirements for compliance. They cite CISPR 11, IEC 1000-4-2, IEC 1000-4-3 and IEC 1000-4-4. Starting in January 1, 1996, these requirements apply to all medical equipment, and will do so until superseded by IEC 601-1, as discussed below.

Product Specific Standards

EN 50024

Also known as CISPR 24, EN 50024 is a draft generic immunity standard for ITE. When enacted, it is expected to use the IEC 1000 series discussed above. While this is not directed at medical devices, we expect that our medical device designers will encounter it someday.

EN 60601. Medical Electrical Equipment

EN 60601 is intended to cite specific safety levels for various types of medical equipment. It is drawn from a subset of IEC 601 (IEC 601-1-2). It becomes mandatory in 1998 (until that time, the generic standards may still be used). As above, the basic standards are cited, but modifications are made, as for operation in the presence of extremely noisy equipment, such as an electrosurgical unit (ESU).

It also cites CISPR 11 and the IEC 1000-4-X series, but is modified for specific applications. The immunity levels are that of IEC 1000-2, 3, and 4, as modified. Note that IEC 1000-4-5, not cited in the generic standards mentioned above, is added to the list.

The ESD test is 3 kV contact discharge to metallic parts and 8 kV air discharge for recessed metal parts. This corresponds roughly to level three in IEC 1000-4-3. As we mentioned earlier, we consider this to be inadequate for use in the U.S.

The RFI test is to level three (3 V/m) as defined in IEC 1000-4-3, except that the test frequency range is expected to be changed to 80 MHz to 1 GHz, as conducted tests will be used for the lower frequencies.

The EFT test selected is level 3 of IEC 1000-4-4, which specifies a 1 kV power line burst for plug connected devices and 0.5 kV on signal lines. This is really a mild test, but will still have to be accommodated. IEC 1000-4-5 testing is also specified at 1 kV differential mode and 2 kV common mode at the power line. Again, this is a fairly mild test.

Potpourri

If you are confused with the number of regulations and their interaction, you have company. To add to the confusion, the EU has its own numbering scheme, and the IEC is busy changing theirs. Table 2-5 shows the equivalencies. The issue has been clouded by the fact that some of these standards have had more than one revision in circulation. Remember that the real governing force, the EU, issuing regulations identified by the prefix EN (preliminary issues are preceded by the letters *pr*), may have adopted a version that is currently in draft and that may be further modified. Nevertheless, the table does serve to sort things out. You may also encounter individual country laws, which will have the same requirements.

U.S. REQUIREMENTS

Unlike the EU, the U.S. medical EMI requirements are not official—but there are de facto requirements. Eventually, we hope to see regulations in common with the EU. Let us look at the various existing standards.

Table 2-5. Standard Equivalencies

ESD	IEC 801-2	IEC 1000-4-2	EN 61000-4-2
RFI	IEC 801-3	IEC 1000-4-3	EN 61000-4-3
EFT	IEC 801-4	IEC 1000-4-4	EN 61000-4-4
Surge	IEC 801-5	IEC 1000-4-5	EN 61000-4-5
Conducted RF	IEC 801-6	IEC 1000-4-6	EN 61000-4-6
Emissions	CISPR 11	EN 55011	
Medical Devices	IEC 601-1-2	EN 60601	
Harmonics	IEC 555	EN 60555	
ITE Immunity	CISPR 24	EN 55024	

Emission Standards

As was mentioned previously, the FCC exempts medical equipment from emission testing, therefore, there are no standards, voluntary or otherwise. The FCC has recently began to accept testing to CISPR limits as valid in lieu of current FCC requirements (Part 15 and 18), and we expect that ultimately everyone will have the same standards.

MDS 201

MDS 201 was prepared for the FDA a number of years ago, but is no longer referenced. Nevertheless, it does contain some useful insights into interference in hospital environments.

FDA EMC Reviewer Guidance for Premarket Notification Submissions

The Reviewer's Guide is one of those regulation holes. There is no official FDA guide, but if you ask the question, they will cite this document. Initially, it was drafted for apnea monitors, then extended to respiratory devices, and finally, with the most recent revision (November 1993), extended to patient connected devices. The FDA has a great deal flexibility when working with this draft document, which would be lost if passed into law. A standard, entitled "Performance Standard for the Infant Apnea Monitor" has been proposed at law, and was published in the Federal Register in February,

1995. We expect significant changes before it becomes law, including expansion to cover more than just apnea monitors.

Generally, the FDA is adopting the immunity portion of IEC 601-1-2, but they are taking a stiffer pose on testing, and have added some requirements, notably the conducted RFI test and the magnetic field test. The conducted RFI test is a bulk current injection on all cables. The test is taken from MIL-STD 461D, test CS114. The actual level cited is quite tough, and is not consistent with the cited 3 V/m test for radiated immunity (the FDA is pressing for higher radiated immunity levels, too). ESD and EFT tests are essentially the same as with IEC 601, and the same comments apply: we think the ESD test is way too low. Magnetic field tests are taken from MIL-STD-461D, test RS101, and is substantial. Most electronics is not affected by low frequency magnetic fields, but CRTs are major victims. The FDA also includes a test for power line variations which we expect will also be adopted by the EU in the future.

The FDA recently flexed their muscles when they discovered unusual susceptibility to powered wheel chairs, and is now calling for a 20 V/m immunity level.

Pacemaker Standard (FDA Contract # 223-74-5083, 1975)

The pacemaker standard test requirement is for a field strength of 200 V/m with 100 percent modulation. Some vendors test to proprietary higher levels as a precaution. Specified test frequency is 450 MHz. This is a severe test, but the implant is inherently easy to shield. There have been recent reports of problems encountered by patients using cellular phones, and this standard may be modified in the near future.

SUMMARY

- Medical device regulations are rapidly evolving. The EU has been spearheading the EMI regulations, but the U.S. has taken a key role in medical EMI requirements
- RFI test requirement of 3 V/m is reasonable for most applications, considering the probable threat, but does pose a significant challenge to sensitive analog circuits, especially for patient connected devices.
- ESD requirements of 3 kV/8 kV (contact discharge/air discharge) are too low, in our opinion. We advise our clients to

strive for 8 kV/15 kV for applications in the U.S. ESD affects mostly digital devices.

- Electrical Fast Transient and surge tests of 2 kV is reasonable, but we think that the signal line test is not appropriate for most medical equipment.
- You may be required to test for several categories, depending on your equipment mix.

3

THE PHYSICS OF EMI

Many in the electronics world feel that EMI is "black magic." Sometimes, you discover a solution to an EMI problem that works well for a specific device, but the next time you try the same fix, it does not work, and, in fact, it even makes the problem worse! At these times EMI seems to defy logic, science, and even simple reason; nevertheless, there are rules of physics that do apply.

The mystification of EMI arises from two sources. First, solutions can be obtained analytically only for the simplest geometries, such as a wire above a plane or a circular loop of wire. Whenever a device deviates from these easily described geometries, EMI problems become unsolvable using standard circuit or field analysis techniques.

Second, there are almost always two or more interference problems existing in concert. There may be multiple sources of interference, multiple receptors, and multiple paths connecting the interference sources to the receptors. Moreover, the connecting paths may be either conducted or radiated, or a combination of the two (e.g., radiated interference is often intercepted by cables and then conducted into electronic circuits). All of these different variations need to be considered before the problems can be controlled or eliminated.

The ability to solve such complex EMI problems is based on an understanding of the underlying physics of the phenomena involved. This chapter focuses on several relevant physical concepts, ranging from unwanted antennas to unexpectedly high-impedance grounds and imperfect shielding.

SOURCE–PATH–RECEPTOR

In chapter 1 we introduced the source-path-receptor concept. In each interference problem there is a source of interference, a receptor of interference, and a path connecting the two. If any of these three ingredients is missing, there is no interference problem. By the same token, if one of these is removed, the interference problem goes away.

But which of these do we attack? If you are concerned about emissions, you would attack the problem at the source. If you are concerned about immunity, you will probably attack the problem at the recipient. In a facility problem you might attack the problem at the path.

In actuality we almost always attack the problem at the path, regardless of the problem, but this usually means close to the source or to the recipient. After all, we cannot truly eliminate the problem at the source unless we are able to somehow prevent the interference from being generated in the first place–usually, we have to block the path very close to the source. Similarly, we cannot usually attack the problem at the recipient, before energy gets there. At high frequencies the path is usually a combination of conducted and field coupling.

THE NATURE OF INTERFERENCE

In order to work with interference, it is necessary to understand some of the physics behind interference. Often, Maxwell's equations are brought out as an aid to understanding interference, but unless you are a real bug on theory, you do not really understand Maxwell's equations all that well. But maybe you understand things a little better than you thought–we believe you can understand a lot of the nature of EMI starting with a basic knowledge of physics.

We say that all interference problems are the same, providing scale factors are taken into account. These scale factors are frequency, amplitude, time, impedance, and dimensions–FAT-ID. Whenever you look at an EMI problem, you need to review these five factors. When you do, you will find that a lot of mystery goes out of EMI. A lot of old lore regarding certain practices makes sense in certain contexts, but not in others. A good example is the single-point ground, which makes good sense when applied to low frequency interference, but is completely inappropriate at radio frequencies, which is where many of the EMI problems lie. But many engineers blindly apply the single-point ground everywhere, often creating a problem rather than solving one.

INTERFERENCE IS A CURRENT

Modern designers tend to think of current as that which results from applying a voltage across an impedance. But from an interference standpoint, it is more instructional to consider a voltage as that which results from applying a current to an impedance. This means looking at circuits from a Norton equivalent rather than a Thevenin equivalent. In doing so many EMI questions answer themselves.

The reason is this: Current always makes a round trip, following one or more paths. And you want to steer this interference current to best advantage. This means that you want to provide a low impedance current path back to the original source of interference (if possible) or to divert interference currents away from the load. By the same token you want a high impedance path to the load. So when looking at the possible paths, look at how the current might divide as a result of the various impedances along the various paths.

FAT-ID

When considering FAT-ID, we need to correlate and connect these factors as much as possible, in order to understand the magnitudes involved. Accurate analysis has proven very difficult, but we can give some rules of thumb–not accurately, but good enough to put you in the ballpark. In fact, we like to say that any EMI analysis that is within an order of magnitude (20 dB) is pretty good. Thus, if your quick and dirty analysis shows you are within an order of magnitude of a problem, you should be prepared for the eventuality that a problem may occur. Often, however, we find that the engineer has absolutely no idea of what magnitudes are involved, hence, some gross approximations are appropriate. So let us look at these scale factors and try to correlate them to the real world.

Frequency and Time

Logic design engineers tend to think in the time domain. Interference, however, is better handled in the frequency domain: filters, shielding materials, gasketing effectiveness, and passive components are all specified in terms of frequency domain. In fact, you do not stand a chance of analyzing your problem in the time domain.

Most engineers have encountered the situation where they have driven a fast rise time signal down a signal line only to find a much slower rise time at the receiving end. The approach is generally to drive the signal harder in an attempt to get it to the far end. Did you ever stop to think what happened to that leading edge? It did not get to the end–where did it go? In well-shielded coaxial cable the losses are entirely contained within the cable–the fast edge rate is lost by shunt conductivity of the dielectric between the inner and outer conductor, and the current inevitably returns to the source. But most cases do not use coaxial cable, much less a tight braid. Accordingly, the lost current goes somewhere outside of its intended path–coupling to adjacent circuits or to nearby structural members. This is known as interference. And this is much more readily understood in the frequency domain.

Unfortunately, much of our design work is done in the time domain. So a ready method is needed to convert between the time domain and the frequency domain. This method is Fourier analysis. Fourier (a French mathematician, circa 1800) proved that any periodic waveform can be resolved into an infinite series of sine waves, each an integral multiple (or harmonics) of the fundamental frequency. Thus, you would expect to see components at f, 2f, 3f, and so on in generally decreasing amplitudes (Figure 3-1). Not all frequency components will be present–symmetrical waves have only odd harmonics. The precise makeup of the harmonics can be determined by mathematical analysis (called a Fourier transform), providing that the wave can be described, or can be easily measured using a spectrum analyzer.

A typical series, such as for a trapezoid wave, looks like as in Figure 3-2. This drawing, shown in linear scales, has several interesting points. First, some harmonics are much lower in amplitude than others, some are zero, but the peak levels follow a definite pattern. We are not interested in the low amplitude contributions (they are not contributors to EMI problems), so we will concentrate on the high contributors, which can be described by an envelope. Similarly, we are not really interested in phase, so we will use absolute values. Finally, we find linear scales inconvenient, so we will use log-log scales. The result, also shown in Figure 3-2, is an envelop plot of the spectral energy of the trapezoid wave.

There are three elements in this plot, two of which are of particular significance to EMI analysis. First, the envelope is presented in the fashion of a Bode plot, with specific break frequencies. The first break frequency comes at $1/\pi t$, where t is pulse width. At frequencies above this break point, the amplitude falls off as 1/f (f is

Figure 3-1. Harmonic Components of a Periodic Wave

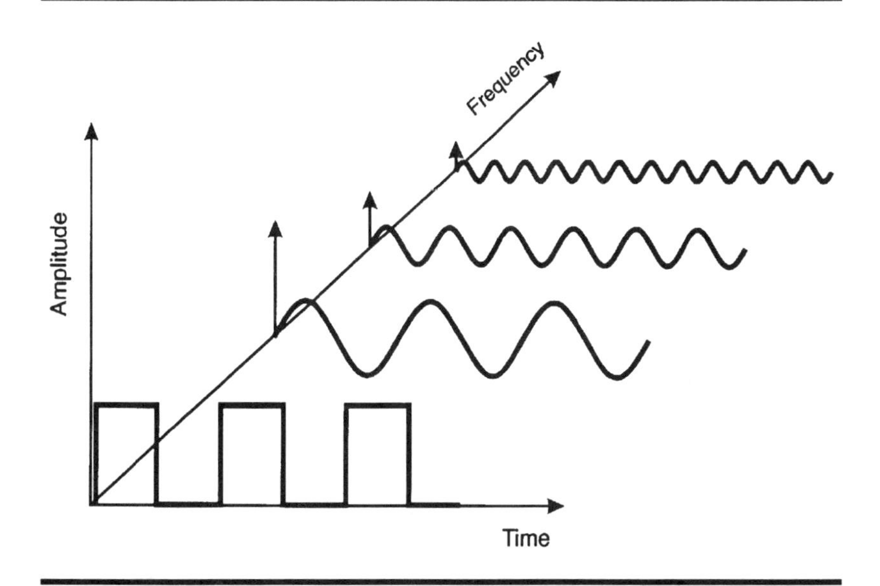

frequency) or at -20 dB/decade, the same as with a plot of a single pole filter. The second break point is at $f = 1/\pi t_r$, where t_r is rise time. At frequencies above this the amplitude falls of as $1/f^2$, or at -40 dB/decade. The spectral spacing is determined by the period and does not impact the shape of the envelope.

Thus, the key issue is that the plot is essentially completely characterized by two key parameters, the pulse width and the rise time (or fall time, whichever is fastest) So we do not need to do a detailed analysis–we just need to know these parameters so we can draw the envelope. We call this the extremely fast Fourier transform, or EFFT–a computer is not necessary.

Figure 3-3 shows some typical waveshapes and their transform. The top wave is a square wave with zero rise time. In practice, no wave has zero rise time (although semiconductor designers have fond hopes), but some logic families are pretty fast relative to the system speed. Note that for a square wave, the second break point moves to the right to infinity as rise time approaches zero.

The next wave is that already discussed previously, and is of interest primarily because it describes the spectral content of a computer clock. The last wave is a special case of a triangle wave, actually the limiting case where the pulse width of the trapezoid

Figure 3-2. Fourier Series for Trapezoid Waves

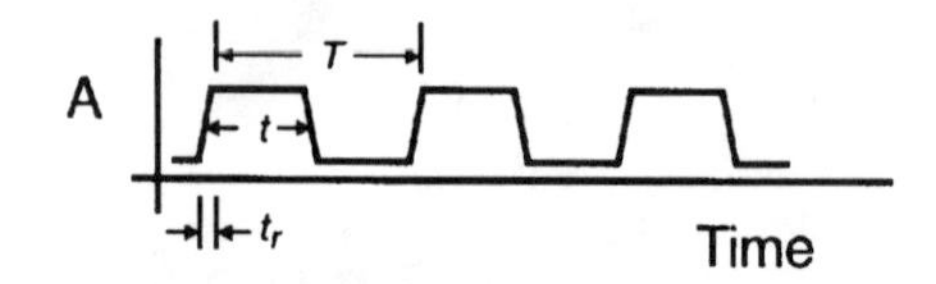

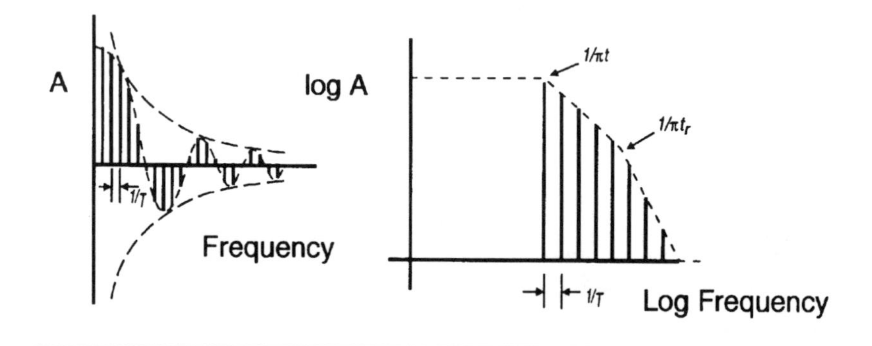

wave becomes small–approximately that of the rise time. In this case the two break points merge, giving a -40 dB/decade fall off. This wave is also of interest, because it approximates a spike of current drawn every time a CMOS clock circuit switches.

These envelopes are of interest, not only because they describe the spectral content of a wave, but they can also be applied to other Bode or frequency plots, such as found in filters and shielding, to arrive at an effective composite characteristic.

As will be discussed later in this chapter, the efficiency of a short antenna (any short wire) increases linearly with frequency, up until about 1/4 of a wavelength, where the antenna is at maximum efficiency. Above this frequency the efficiency depends on standing wave conditions, but the maximum approximates that of 1/4 wave. It is convenient to plot antenna efficiency as an envelope as in Figure 3-4. Also in Figure 3-4 is the envelope of three waves, all of the same frequency, but with different rise times. We can see that as the amplitude of harmonics decreases at –20 dB/decade (1/f) the antenna becomes more efficient at a rate of +20 dB/decade. If we apply the spectral content of the wave to the antenna (using dB scale, the plots add), it can be seen these effects cancel, so that the effective amplitude of radiation remains constant (independent of

Figure 3-3. Fourier Series for Common Waves

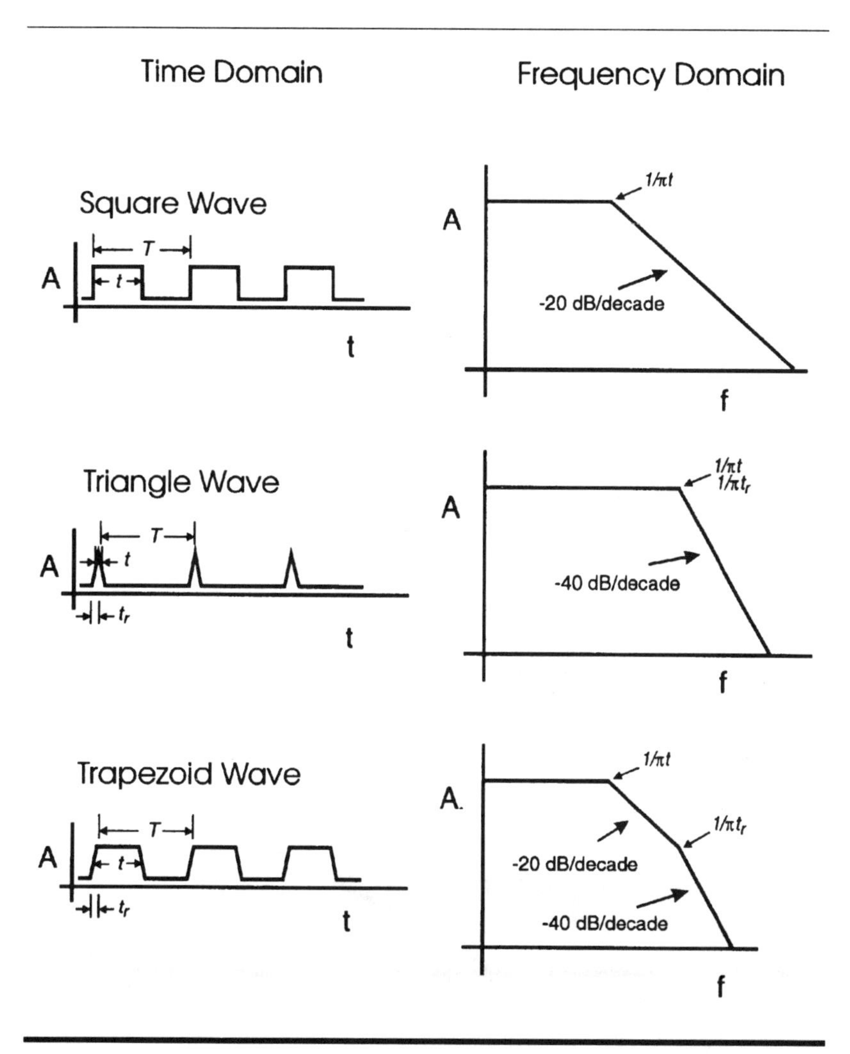

frequency) until the second break point is reached. Above that frequency the energy level starts to diminish.

Although this description is given as an emission issue, the same concept applies to immunity issues: High frequency energy couples very efficiently; therefore, the higher the frequency content of the incident wave, the more gets coupled into your sensitive circuit.

Figure 3-4. Harmonic Emissions for Several Logic Speeds

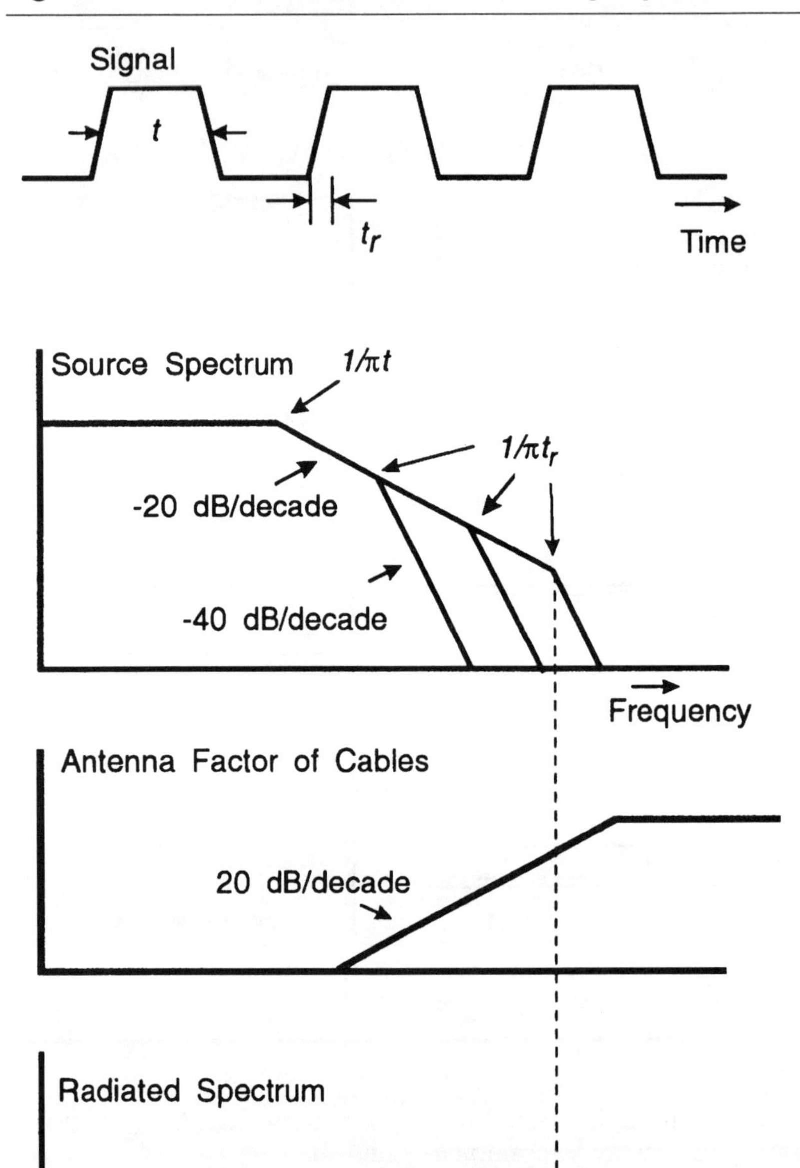

Therefore, the rise time is the single most important factor in energy coupling. Note, that although there are many more transitions per second with periodic pulse coupling than with single pulse coupling, the energy per cycle is identical. Thus, emission levels are directly proportional to frequency, and reducing frequency will reduce average emissions; but the transition energy is still the same, and this is the issue for immunity and cross-talk in digital systems.

Returning to the Fourier transform, note that the envelope of the wave is independent of the frequency. As the frequency decreases (keeping the pulse width the same), the spectral lines become closer spaced, but the envelope remains the same. If the frequency falls to zero, which represents a transient, the spectral lines become a continuum, but the envelope remains the same.

Thus, we can treat transient waves exactly the same as periodic waves. Figure 3-5 shows two common transients of interest, ESD and lightning. ESD has a rise time of about 1 ns or an equivalent frequency of 300 MHz. Lightning has a rise time of about 1 μs, or an equivalent frequency of 300 kHz. Thus, as we look at circuit parameters in the following paragraphs, we will note that the critical frequency, as defined by the rise time, will always be an important factor. Now we have the means to correlate time with frequency, and we will subsequently do most of our analysis in the frequency domain.

Frequency and Impedance

Let us apply the frequency content of any type of interference, whether continuous or transient, to impedances. As is well known, the impedance of inductors and capacitors is a function of frequency. Parasitic inductance and capacitance both play a significant role in interference. Series inductance in the wire blocks high frequency components. Stray shunt capacitance provides alternate paths for the current.

Inductance and capacitance exist not only as functional components bought as such, but they are also in existence as stray elements in these same components and in structural members. We have several cases where impedance is a concern: components, ground paths, and other structural elements. Let us look at each of these in turn.

In the real world the behavior of electronic components can deviate significantly from idealized models. Capacitors, resistors, and inductors all exhibit parasitic behavior. Second and third order parasitic effects are generally ignored during the product design phase;

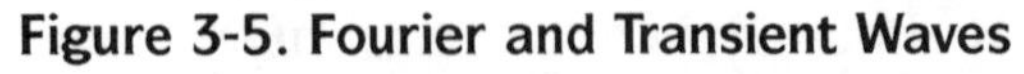

Figure 3-5. Fourier and Transient Waves

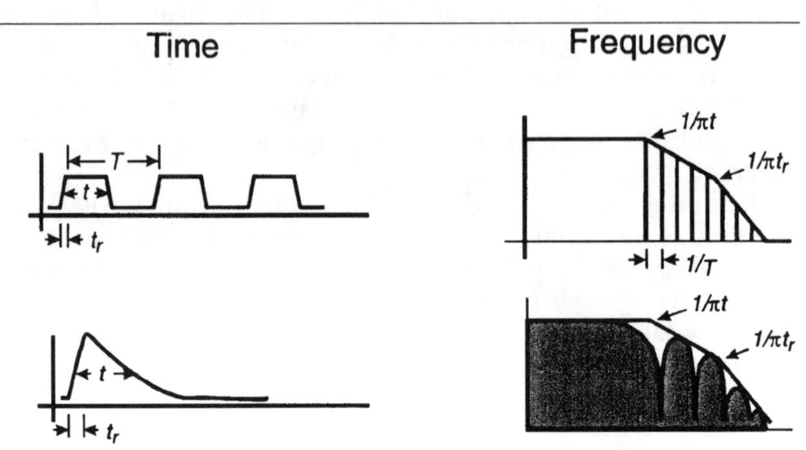

Examples

Type	Pulse Width	1/πt	Pulse Rise	$1/\pi t_r$
ESD	30 ns	11 MHz	1 ns	320 MHz
Lightning	50 ms	6.3 kHz	1.2 μs	265 kHz

however, parasitic effects become major factors in component behavior at high frequencies. A component that is well behaved at the clock frequency may become a completely different circuit at the 10th harmonic. In fact, most components degrade at much lower frequencies than commonly expected. The behavioral characteristics of components will be discussed in detail in chapter 4, but briefly, the following rules of thumb apply:

A wire may be assumed to be a conductor at low frequencies, but becomes noticeably inductive at higher audio frequencies. Wire inductance is approximately 8 nH/cm, a value that varies little with diameter or conductor shape. Thus, even a run of a few inches on a circuit board can become significant. At still higher frequencies wire acts as an antenna, becoming significant at 1/20 of a wavelength and reaching peak efficiency at 1/4 to 1/2 of a wavelength. For example, a 10 cm wire will become a significant antenna at 150 MHz and a very efficient antenna at about 800 MHz. As another example, a 1 cm standoff, used on a circuit board has about 8 μH of inductance (L). At 300 MHz (ESD frequency), the impedance (Z) of that standoff is

$$Z = 2\pi f L = 15\ \Omega$$

Similarly, capacitance (C) in structural members provides alternate current paths for high frequency currents. Assuming a circuit board about 30 cm square (approximately the size of a personal computer motherboard), spaced one cm above the chassis has a capacitance of

$$C = \varepsilon_0 A/d$$

where ε_0 is 8.85×10^{-12} F/m. If $C = 80$ pF, then

$$Z = 1/2\pi f C = 6\ \Omega \text{ at } 300\ \text{MHz}$$

As can be seen in this example, the impedance of the ground strap and circuit board is such that much of ESD current would flow down the capacitive path rather than the intended ground path. Stray capacitance and inductance paths become extremely critical in medical electronics, where leakage currents are limited power supplies (as will be seen in chapter 7).

Your purchased components degrade also. A capacitor can be modeled as a true capacitor in series with an inductor (series resistance is usually neglected for high frequency capacitors), as determined by the lead length internal and external to the component. A series resonant circuit is inductive above resonance.

Similarly, an inductor has a distributed interwinding capacitance that dominates the inductor at higher frequencies. Most purchased air-core inductors will resonate at frequencies from a few MHz to about 50 MHz, are infinite impedance at resonance, and are capacitive above resonance.

Finally, the actual characteristics of resistors with lead and spiral inductance, and shunted by a small capacitance, will vary widely. Generally, for lower resistor values (say, less than 2 kΩ) they become capacitive at lower frequencies, then become inductive at high frequencies.

In sum, in operation, all electronic components will exhibit significant deviations from their ideal behavior at the frequency levels where EMI problems are generally encountered.

Dimensions and Frequency

Physical dimensions are a significant factor relative to a wavelength. This is the point where circuit analysis can no longer assume lumped circuit parameters.

This situation is well known to designers of high speed circuits, where the speed of light is a significant part of the propagation time from driver to receiver. When line length becomes long relative to a wavelength, or, in time domain terms, when the rise time becomes shorter than the propagation delay time, then the line becomes a transmission line, and must be terminated in its characteristic impedance to eliminate signal reflections. While this practice is primarily to preserve signal integrity, it also helps to control EMI.

But for EMI purposes, our interest is more in the dimensions relative to an antenna. As was mentioned earlier, when the length of a wire approaches 1/4 of a wavelength, then it looks like an efficient antenna. We also mentioned that wire lengths of 1/20 of a wavelength start to look like an efficient antenna, and we will take this same criteria to define the approximate boundary between a lumped system and a distributed system. Thus, at frequencies less than 1/20 of a wavelength, the circuit may be analyzed as a lumped circuit. Above that frequency distributed effects start to become significant.

Antennas

Power lines, signal cables, and openings in an equipment enclosure can inadvertently act as antennas, conveying energy into or out of a circuit. Any conductor can be an antenna, but dipole, loop, and slot antennas are the most common types encountered in EMI. Functionally, all of these operate with approximately equal efficiency. Let us take a look at antennas from a physical standpoint.

Figure 3-6 shows a small dipole and a small loop antenna. If we assume the antenna element is very short relative to a wavelength, we can readily see that the dipole antenna is a high impedance circuit (almost open circuit) and the loop antenna is a low impedance circuit (almost a short circuit). Thus, we would expect the electric field to dominate in close proximity to the short dipole, and we would expect the magnetic field to dominate in close proximity to the small loop.

That this is true is seen in field solutions of these antennas. The field solutions of antennas generate a variety of electromagnetic field components, which can be categorized as far field and near field. Far-field (or radiated) components follow the $1/r$ law (where r is the distance from the source), while near-field components are very short range, following reciprocal square and cube laws. The latter are undetectable at any distance from the source, but dominate very near the source and are the principal contributors to cross talk.

Figure 3-6. Small Dipole and Loop Antennas

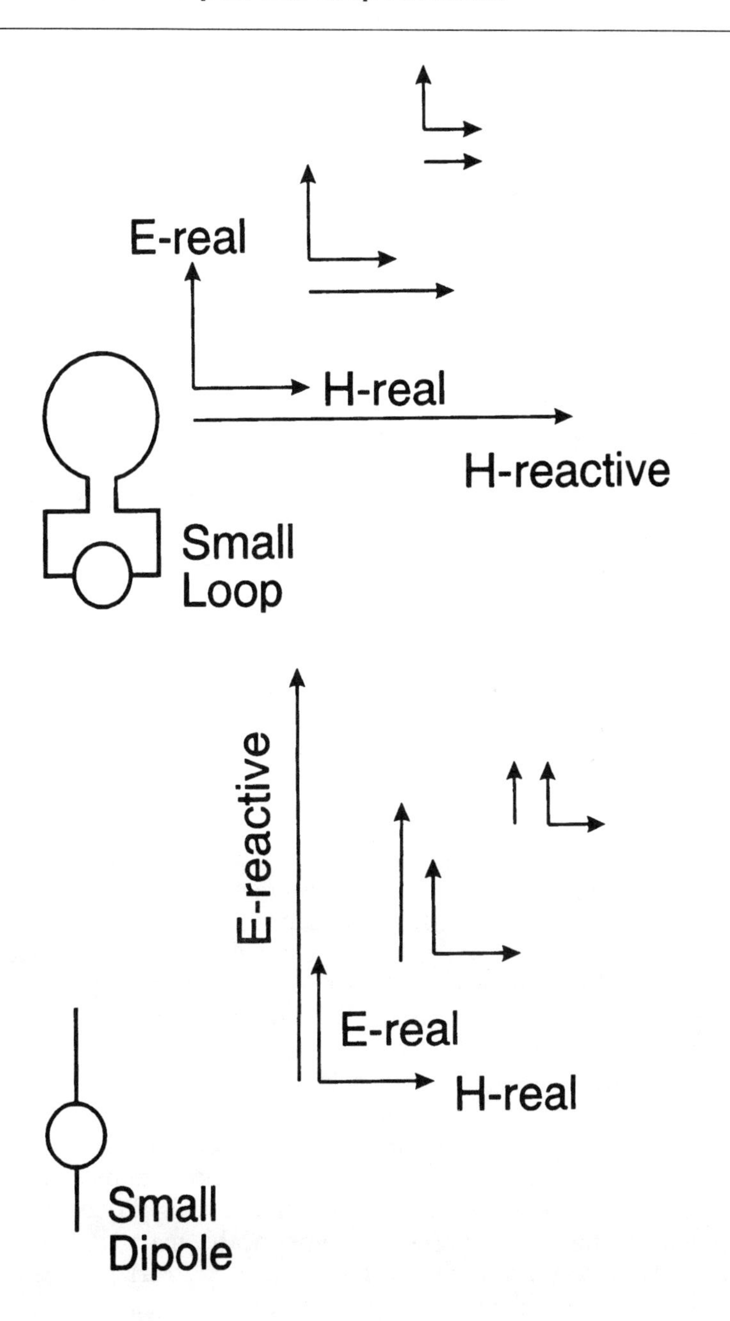

The effectiveness of a short dipole antenna (<1/4 of a wavelength) is proportional to the product of the frequency, the current amplitude, and the length of the antenna. The effectiveness of a small loop antenna (<1/4 of a wavelength) is proportional to the product of the square of the frequency, the current amplitude, and the loop area. Of the three factors the physical dimension (length or loop area) is the most readily controlled.

As Figure 3-7 indicates, computing the two antenna types' wave impedance (the quotient of the electric field divided by the magnetic field at any point in space) reveals that in the near-field area, a dipole generates a high impedance wave and a loop generates a low impedance wave. As distance from the source increases, these impedance values quickly converge to a common value of 377 ohms. As will be discussed in chapter 9, wave impedance has a major impact on shielding effectiveness.

The slot antenna is the analog of the dipole antenna, where the conductor is replaced by an insulator and the insulator is replaced by a conductor. The effect is shown in Figure 3-8, where the laminar flow of the current on the conductor is diverted by the presence of the slot. Actually, the criteria of selecting one over the other is from dimensional considerations. At lower frequencies wire members such as used in dipole or loop antennas are far more feasible than using a slot antenna. But as frequency increases, the desired length of the antenna element decreases, until it becomes very difficult to tap for signal purposes (how would you center feed a 3 cm dipole?), so higher frequencies use slot antennas.

Shielding Effectiveness

The effectiveness of EMI shielding devices results from two mechanisms: reflection and absorption. Reflection is a surface phenomenon, which depends on the impedance mismatch between the incident wave and the shield boundary. A good conductive surface will reflect a high impedance wave (such as a 377 ohm plane wave), but reflection is very low when the wave impedance is low (i.e., a magnetic field). Absorption is a bulk effect and is achieved effectively only with permeable materials.

Most shielding problems fall into one of two categories: low frequency magnetic fields and high frequency electromagnetic fields. Low frequency magnetic fields, such as those found in power supplies, including 60 Hz and chopper frequencies, do not reflect effectively and must be shielded with a permeable material, the thicker the better. In contrast, high frequency waves (>1 MHz) are easily reflected by a thin shield of any good conductive material, including

Figure 3-7. Wave Impedance Near Source

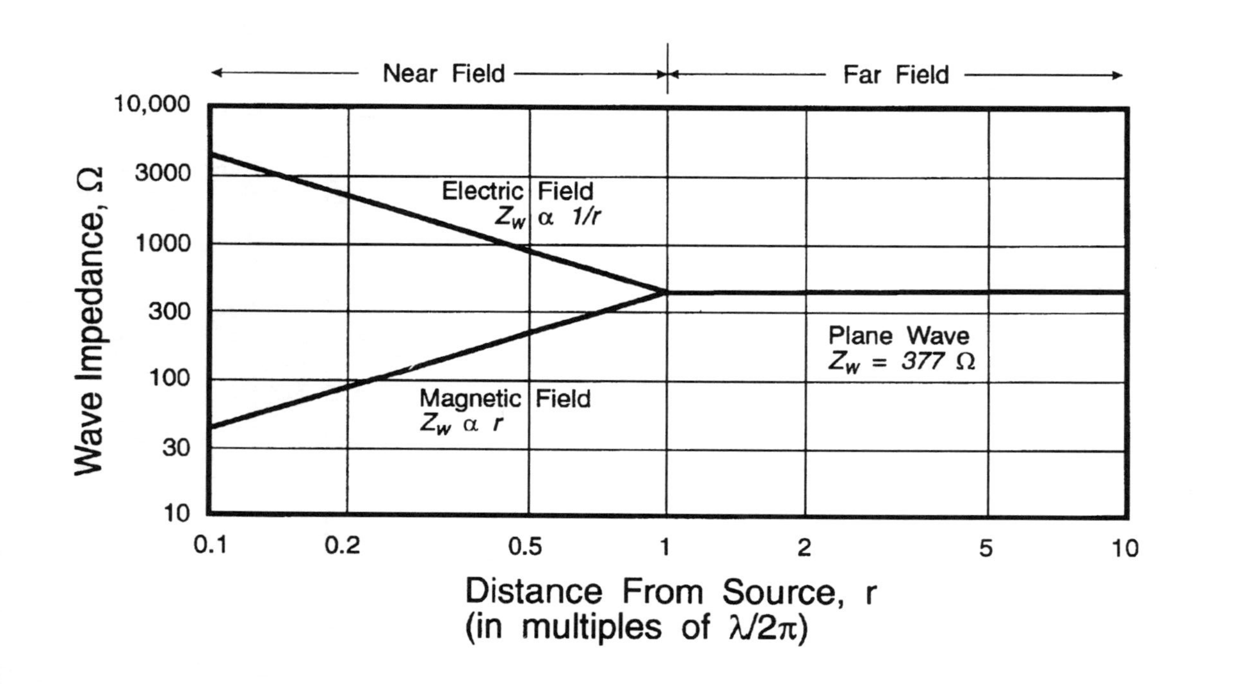

Figure 3-8. Slot Antenna

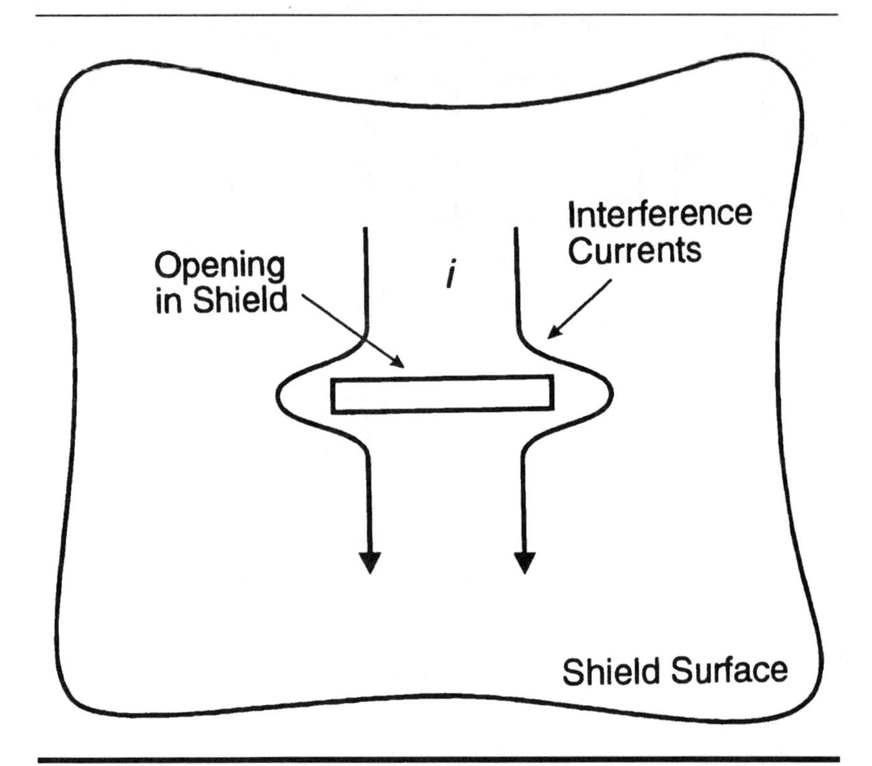

coated plastics. Thickness is not important. In fact, most shielding failures occur, not because of a deficiency in the shielding effectiveness of the material, but with openings and penetrations in the shield.

Amplitude

Amplitude impact is self-explanatory: the higher the amplitude, the more interference you have. But what amplitudes might we expect? We can take some simple cases to show approximately what amplitude we might expect for certain situations. And some of the above factors (frequency and dimensions) become significant factors in such computations.

Field Strength

For a given power source what field strengths may we expect from a specific distance? We can use the relationship

$$\boldsymbol{E} = \frac{5.5\sqrt{P}}{R}$$

where P is effective radiated power in watts, R is distance in meters, and E is field strength in volts/meter. This is arrived at by assuming an isotropic radiator (uniform in all directions), computing power density at a given distance, and assuming the wave impedance is 377 Ω. Actually, an antenna factor should be applied to P, but, in practice, the estimate will be crude enough that this factor can be ignored.

As an example, what is the electric field strength from a 1-watt handheld radio held 1 meter away?

$$\boldsymbol{E} = \frac{5.5\sqrt{1}}{1} = 5.5 \text{ V/m}$$

This is above the IEC 1000-4-3 requirement of 3 V/m.

Field Strength and Voltage

What is the relationship between electric field strength and voltage on a wire antenna?

For 1/4 of a wavelength antenna

$$E = v\mathrm{f}/33 \text{ or } v = 33E/\mathrm{f}$$

where E is the electric field strength in V/m, v is the induced voltage (in volts) on the wire (assuming high impedance load), and f is the frequency in MHz. For rough estimates a long cable of undetermined length exposed to a wide range of frequencies can be expected to look like a quarter wave antenna at some frequencies. Accordingly, we use this rule to compute the approximate worst case voltage that might appear on a wire randomly placed (for many frequencies, it will be much less).

As an example, what is the induced voltage for an incident wave of 3 V/m onto a long patient cable at 30 MHz?

$$v = 33 \times 3/30 = 3 \text{ V}$$

Thus, your analog circuit can be expected to be assaulted by a 3-volt RF signal.

Field Strength and Current

What is the relationship between electric field strength at a specific distance and the current on a wire? Use the following relationship:

$$E = 2\pi \times I \times \ell \times f/10 \times R$$

where I is current in amps, ℓ is length of wire in meters, f is frequency in MHz, and R is distance from wire in meters.

For example, measure the current on a cable, and extrapolate to see if it is significant compared with CISPR 11 limits. Suppose I is measured at 20 μA, ℓ is 3-meter cable, f is 100 MHz, and distance is 10 meters.

$$E = 2\pi \times 20 \times 10^{-6} \times 100/10 \times 10 = 94\ \mu V/m$$

This is just below the CISPR limit of 100 μV/m. As our example is only an approximation, we would see this level as a potential emission problem.

The Threat and Physics

Let us connect up the principal threats (emissions, RFI, ESD, power disturbances, self-compatibility) to FAT-ID. The physics of the threat dictates the nature of the remedy. We would not apply low frequency remedies to high frequency threats, not because the physics are different, but because the scale factors are different.

The frequency of the threat, whether emissions or reception, gives a good indication of the path the interference will take. Recalling some rules of thumb given above, a 1/20 of a wavelength conductor is generally considered to be the boundary between a good antenna and a poor antenna. Hence, low frequency interference does not propagate well by radiation.

We mentioned that conductor inductance is about 8 nH/cm, which means that high frequencies do not propagate well down a wire. So low frequencies propagate primarily by conduction and high frequencies propagate primarily by radiation. The regulatory agencies have selected 30 MHz as a boundary above which radiation is tested and below which conduction is tested. This is somewhat arbitrary, because both radiation and conduction will obviously exist side by side, but 30 MHz is a good boundary.

We can draw another distinction in the radiated frequencies. Again, using our 1/20 of a wavelength rule, we find the lower frequencies still have a fairly long wavelength, and the only antennas long enough to provide for such radiation is the cable, either data or power. Slot radiation from enclosures and board radiation from circuit boards is still not very efficient at lower frequencies, because their dimensions are too small for efficient radiation. At about

300 MHz (e.g., ESD frequencies), however, the wavelength has now become small enough that enclosure seams and circuit boards become good antennas.

So we can summarize by saying that the frequency of the threat gives you a good hint as to the path:

- $f < 30$ MHz, interference is primarily conducted
- $30 < f < 300$ MHz, interference is primarily radiated via cables
- $f > 300$ MHz, interference is primarily radiated via cables, enclosure openings, or PCBs (printed circuit boards)

This rule of thumb can be applied to any of the threats, whether transient or continuous wave and for emissions and immunity.

ESD

The physics of ESD needs to be explained a bit, as there is some confusion on what the mechanism is. Human ESD has a 1 ns rise time, which we will call a 300 MHz problem. There are really several mechanisms involved. The first is direct injection as shown in Figure 3-9. This is the harshest test, as energy may be injected directly into a circuit. Usually the signal conductor is insulated during normal operation, but is vulnerable when installing cables, where an installer can easily inject ESD directly into a male connector pin. After cabling the signal line is generally protected from direct ESD, but the ground path may well still be exposed.

Indirect ESD effects can be either electric field or magnetic field coupling. Although ESD implies an electric field situation, magnetic field coupling arising from the high currents is much more commonly a problem. Figure 3-10 shows a case where the ESD current is injected to the ground at the perimeter of the enclosure. If the enclosure is plastic, the magnetic field coupling path to an internal circuit member can be quickly estimated:

$$\mathrm{V} = -\frac{d\theta}{dt} = \frac{A\mu_o}{2\pi\ell} \times \frac{dI}{dt}$$

If A = area = 1 cm^2, ℓ = distance = 10 cm, dI = 10A, dt = 1 ns, then V = 2V

The intercepted loop is sufficient to upset most modern logic.

Figure 3-11 shows a capacitive coupling path. This most often occurs when an enclosure door is inadequately grounded. The

Figure 3-9. ESD Current Injection

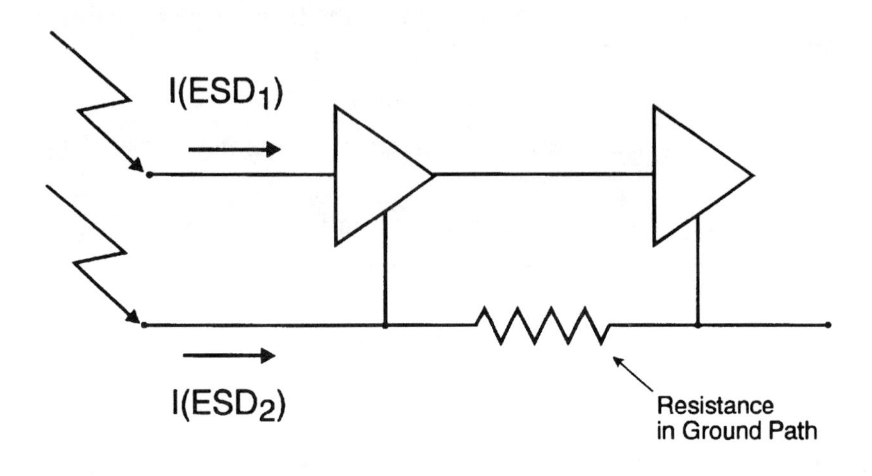

Figure 3-10. Inductive ESD Coupling

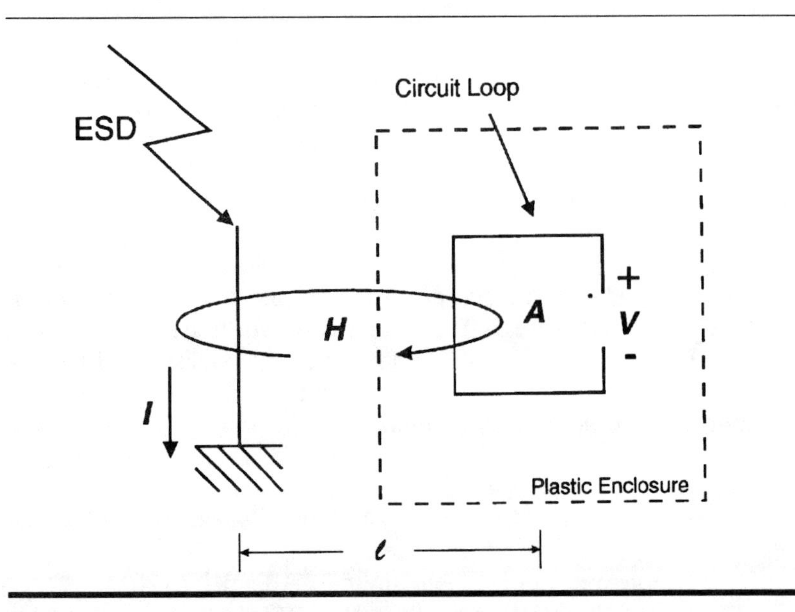

discharge provides a substantial electric field coupling path to internal cables and PCBs.

Figure 3-11. Capacitive ESD Coupling

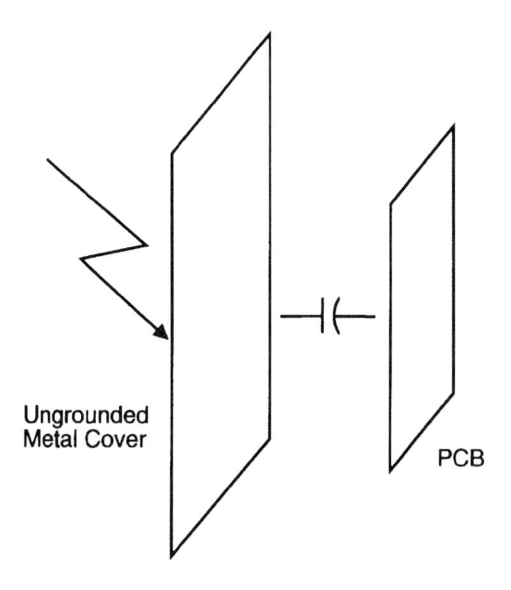

SUMMARY

The following is a checklist of important points made in this chapter:

- Consider EMI to be a current, and remember that current always returns to its source.
- Rise times are a most important parameter in interference considerations. Remember EFFT, which gives an equivalent frequency (f_{eq}) of $1/\pi t_r$ (ESD, 1 ns, 300 MHz).
- Inductance of a wire is approximately 8 nH/cm (5 Ω/cm at 100 MHz).
- Components degrade at high frequencies. Inductors become nonfunctional at frequencies usually well below 100 MHz.
- Wires and openings start to become efficient antennas when the lengths are greater than 1/20 of a wavelength. Thus, a 300 MHz source has wavelength of 1 meter, and 1/20 of a meter is 5 cm.

- Loop areas of signal or power cables are the single most important and controllable element in any design project.

The following chapters will amplify these concepts and apply them to real world electronic device applications.

4

COMPONENTS AND CIRCUITS

Many electronics designers mistakenly assume that the components they select and use will not cause EMI problems. This is particularly true with passive components, which many designers assume are ideal. If only life were so simple in the real world, where the root causes of many EMI problems are due to improper component selection and application.

In this chapter we will look at the subtle ways EMI can affect passive and active components, and give you some guidelines in selecting components with EMI in mind. We will also look at transient protection devices, and show you how to design and implement simple EMI filters.

Most EMI component problems are due to exceptions to the rules. For example, when is a capacitor not a capacitor? When it is an inductor, due to lead length at high frequencies. When is an inductor not an inductor? When it is a capacitor, due to parasitic wire coupling at high frequencies. Even simple wires behave as antennas at high frequencies. The secret to success with components is knowing their limitations, and then designing to accommodate these limitations.

We often refer to these component exceptions as part of the "hidden schematic," as shown in Figure 4-1. It is these unexpected effects, in conjunction with other unexpected parasitic capacitance to adjacent wires that contribute to many EMI problems. The higher the frequency, the more likely the "hidden schematic" will be a factor. It is only by understanding these factors that we can design to prevent their associated EMI problems.

PASSIVE COMPONENTS

In this section we will look at the EMI characteristics of several common components, such as capacitors, inductors, resistors, and

Figure 4-1. Components vs. Frequency

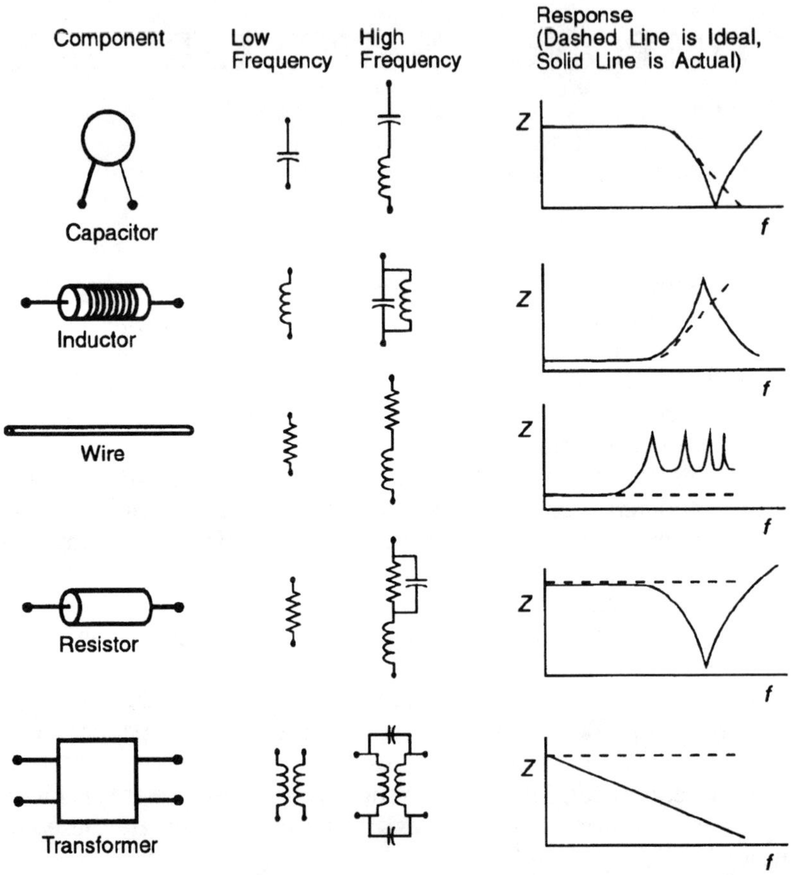

transformers. But first we will look at the most basic of "components," interconnecting wires or traces. We will see that the performance of these components is greatly affected by frequency. Our focus will be on the higher frequencies, where parasitic effects can have serious impact on EMI issues.

Wires and Traces

Although not generally considered a component, the wires and traces that connect components play very important roles when it comes to EMI. Even wires that are considered parts of the components themselves, such as leads on capacitors, affect EMI characteristics. These effects are due to parasitic inductance, parasitic capacitance, and antenna behavior that greatly affect wire impedance with frequency.

This variation with frequency is illustrated in Figure 4-2. At low frequencies the wire is primarily resistive, then it becomes inductive, and finally it becomes a transmission line or antenna. As will be seen later in the chapter on grounding, these impedance changes also have profound impact on grounding strategies, leading us into ground planes, grids, and so on. For now, however, we will concentrate on the characteristics of the common round wire and its cousin, the PCB trace. Let us first look at ***parasitic inductance.*** The impedance (Z) of any wire consists of both resistance (R) and inductance (L) ($Z = R + j\omega L$). For DC and low frequencies resistance is the major factor, so heavy gauge wires or large traces are often used for high current paths. As the frequency increases, however, the inductance becomes more important Somewhere in the range of 1–10 kHz, the inductive reactance exceeds the resistance, so the wire or trace is no longer a low resistance connection, but rather an inductor. Remember, any connection above audio frequencies is inductive, not resistive.

The formula for the inductance of a wire is $L = 12.9 \times l \times \ln(4l/d - 0.75)$, l being length and d being diameter in cm. This is based on the concept of "partial inductances," since technically inductance is

Figure 4-2. Impedance of Wire vs. Frequency

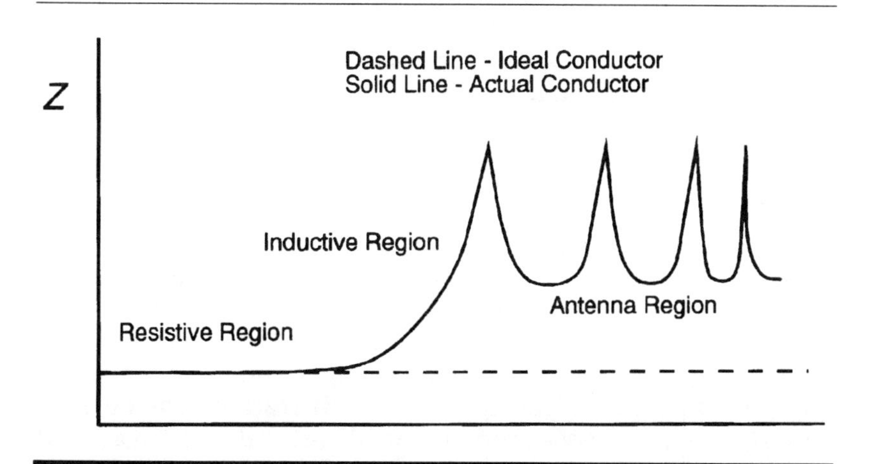

associated with a loop that carries current. In any event the inductance is relatively unaffected by the wire size, since it varies as the logarithm of circumference. This makes sense when you remember that inductance is merely the ability to store a magnetic field. The more surface, the more inductance. This can be boosted, of course, by wrapping the wire in a coil or by adding a permeable material nearby. Both of these increase the magnetic flux, and thus the inductance. The inductance of common wires and traces varies between 6 and 12 nH/cm, so a common rule of thumb is to assume 8 nH/cm for PCB traces and connecting wires.

Next, let us look at the ***transmission line*** and ***antenna effects***. As the frequency continuous to increase, that same piece of wire or trace starts to behave like a transmission line. That is, the voltages and currents are not constant along the length, so we must account for effects like characteristic impedance and reflections. These same wires or traces can also radiate as antennas, or couple energy to nearby wires through crosstalk. Although most antennas are designed to be 1/4 or 1/2 of a wavelength long, a common rule of thumb in the EMI community is to assume that antenna effects occur when lengths reach 1/20 of a wavelength.

> For example, a 10 cm trace has a resistance of 57 mΩ. Assuming 8 nH/cm, the inductive reactance reaches 50 mΩ at 100 kHz. This means that at frequencies greater than about 100 kHz, the trace is inductive, not resistive. That same 10 cm trace is 1/20 of a wavelength long at 150 MHz, so at frequencies above 150 MHz, we assume that same trace is now acting as a pretty good antenna.

We have seen that traces or wires are, in fact, significant components in their own right at surprisingly low frequencies. We will soon see that these inductive effects also play havoc with component leads as well.

Capacitors

Capacitors are widely used in EMI control for power decoupling and filtering. In an ideal capacitor the capacitive reactance decreases linearly with increasing frequency. This is described by the simple formula $X_C = 1/(2\pi f C)$, where X_c is the capacitive reactance in ohms, f is the frequency in Hz, and C is the capacitance in microfarads. Thus, a 10 μF electrolytic capacitor has a reactance of 1.6 Ω at 10 kHz, which decreases to 160 μΩ at 100 MHz. If we need a "short circuit" at 100 MHz, this should certainly do the job.

Unfortunately, we will never see a capacitive reactance that low with our 10 μF capacitor, since the internal inductance and resistance of the electrolytic capacitor will probably limit its useful upper frequency to less than one MHz. Table 4-1 gives some guidelines we use for capacitors.

A second factor that must be considered is the inductance of the leads, which we can estimate at 8 nH/cm. Since this parasitic inductance is in series with the capacitor, a series resonance circuit is formed. At frequencies below the resonance, the capacitor functions as a capacitor, but above the resonant frequency, the capacitor is now an inductor, with its impedance increasing with frequency.

This effect is shown in Figure 4-3, which demonstrates the resonant effects of a 1 cm lead for three values of capacitors. At 10 nF (0.1 μF) the resonance occurs at 5.6 MHz, while at 100 pF the resonance occurs at 178 MHz. Decreasing the lead length by a factor of 2 would raise the resonant frequencies by a factor of the square root of 2.

There are several things worth noting here. First, for high frequency capacitors ($f > 10$ MHz) it is usually the lead length that limits the capacitor's usefulness, not the internal inductance or resistance. *Keep those lead lengths short!* Second, at higher frequencies smaller capacitors sometimes work better than bigger ones. Note than in this example, the 1 nF capacitor outperforms the 10 nF above about 50 MHz, and the 100 pF outperforms the 1 nF above about 100 MHz. It is well worth remembering–next time you suspect poor decoupling at a higher frequency, use a smaller capacitor rather than a bigger one.

Table 4-1. Capacitor Guidelines in Terms of Frequencies

Aluminum electrolytic	100 kHz
Tantalum electrolytic	1 MHz
Hi K Ceramics	5 MHz
Paper	5 MHz
Mylar® film	10 MHz
Polystyrene	500 MHz
Mica	500 MHz
Ceramic	1000 MHz

Figure 4-3. Impedance of Leaded Capacitors vs. Frequency

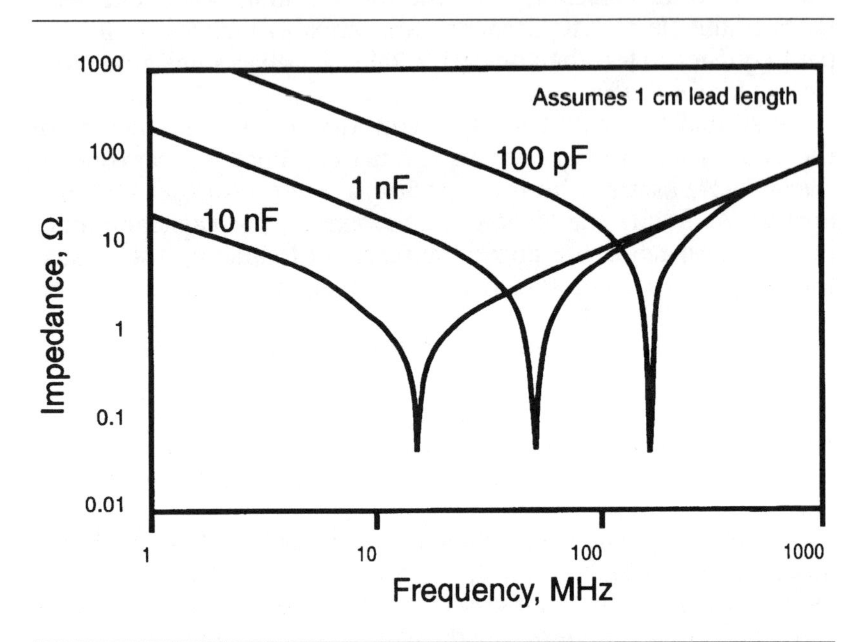

At frequencies above 100 MHz, special capacitor installations may become necessary. For example, surface mount capacitors do not have any leads, and perform well up to 1 GHz. (This assumes they are used on a multilayer board. On a two-layer board the trace inductances prove as devastating as lead inductance.)

The "three-lead" capacitor, shown in Figure 4-4, is popular at high frequencies. In this case the lead inductance is forced "outside" and no longer forms a resonant circuit. In fact, the inductance actually helps in filtering applications, as it forms a small "T" filter. This T filter is often augmented with ferrite beads to increase the inductance, making a very effective and compact filter. Note, however, that the middle lead must be kept short, or the inductance in that lead will again cause resonance problems.

Another high frequency filter is the "feed-through" capacitor. This is often used in connectors or in shield penetrations. In this case the capacitor is coaxial, and the outer lead is solidly attached to the shield or connector frame to minimize inductance. A similar type of high frequency capacitor in connectors is the discoidal array, which also uses the coaxial approach on multiple pins. Like the feed-through capacitor, discoidal capacitors must be bonded to the connector frame through a very low inductance bond, which usually means a full circumferential bond to the frame. These capacitors are shown in Figure 4-5.

We have seen that for high frequencies, you must not only select the right type of capacitor, but you must also pay very close

Figure 4-4. Three-Lead Capacitor

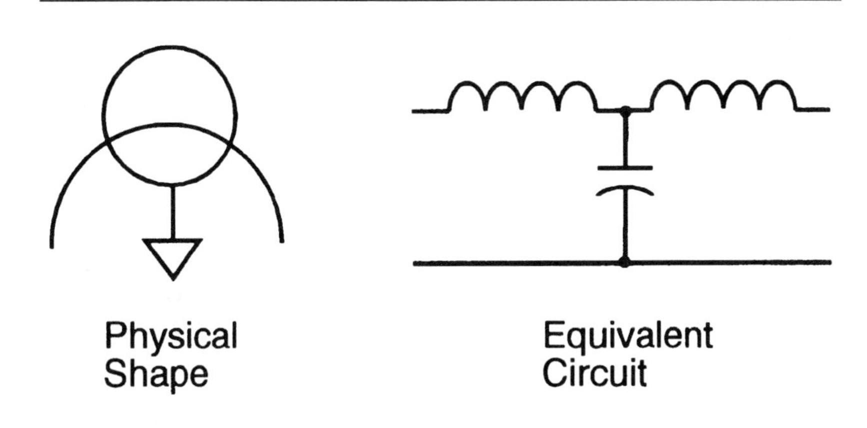

Figure 4-5. Feed-Through Capacitor has Low Inductance

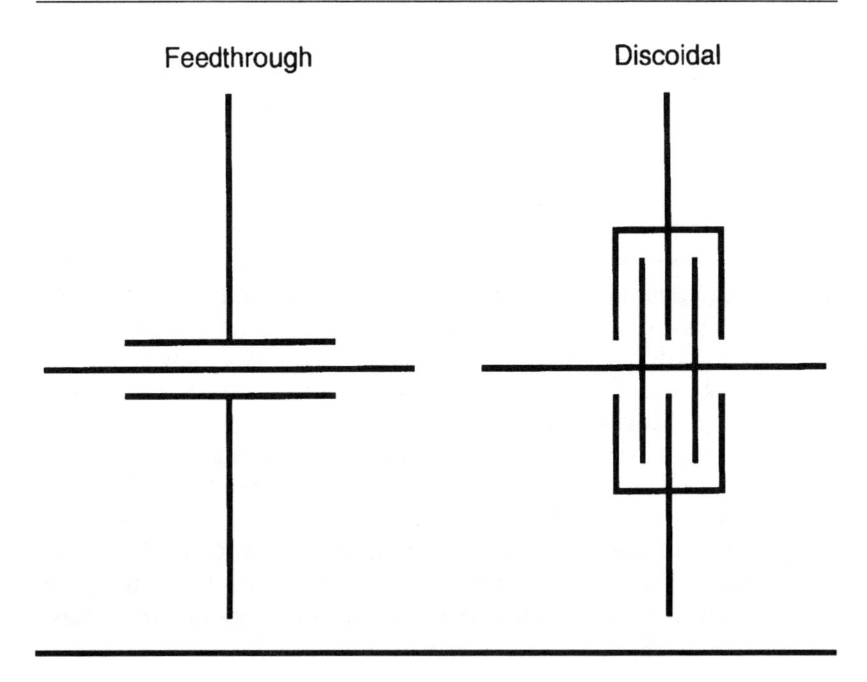

attention to how the capacitors are installed. Left to chance, most capacitors simply will not work at high frequencies.

Inductors

Like capacitors, inductors are also widely used in EMI control for power decoupling and filtering. In an ideal inductor the capacitive reactance decreases linearly with increasing frequency. This is described by the simple formula $X_L = 2\ \pi f L$), where X_L is the inductive reactance in ohms, f is the frequency in Hz, and L is the inductance in henries. Thus, an "ideal" 10 mH choke has a reactance of 628 ohms at 10 kHz, which increases to 6.2 MΩ at 100 MHz. If we need a "open circuit" at 100 MHz, this should certainly do the job.

Unfortunately, we will never see an inductive reactance that high with a 10 mH choke, since the parasitic capacitance between the windings will limit its useful upper frequency to less than 1 MHz. Even small wire wound inductors in the 1–10 μH range start to become ineffective in the 10–50 MHz range due to parasitic capacitance.

So what do we do when we need inductance at higher frequencies? We turn to ferrites, which are alloys of iron/magnesium or iron/nickel. These materials have high permeabilities, that provide for higher impedance with a minimum of capacitance killing turns. Furthermore, most ferrites become lossy as frequency increases, which is actually a benefit for most EMI applications. The net result is a low-Q element that absorbs energy, instead of ringing or reflecting it.

As we are fond of saying, "ferrites are your friends" for EMI and ESD suppression. For years materials engineers and scientists worked hard to reduce the high frequency losses in these materials. Some years ago ferrite manufactures realized that these losses could actually be advantageous for EMI purposes. Instead of trying to reduce the losses, they actually began to formulate materials that were optimized for their loss characteristics. The key issue is that adverse energy could now be absorbed and converted to heat instead of reflecting and causing trouble elsewhere.

The results have been very fruitful for EMI and ESD applications. The most common type of ferrite for EMI (NiZn) becomes lossy above about 10 MHz, and reaches a peak at about 100 MHz. Below 10 MHz the inductance predominates, but for EMI/ESD applications, we prefer to operate them in their lossy range anyway. In a sense they are "frequency-dependent" resistors. Some ferrites have low volume resistivity (e.g., MnZn) and must be isolated or insulated.

Figure 4-6 shows the composite impedance (reactive and resistive) of two popular EMI ferrite formulations. The lower frequency material has a relative permeability of 2500, while the higher frequency formulation has a relative permeability of 850. In fact, most manufacturers specify the "equivalent resistance" at 25 or 100 MHz as a key design parameter. For most EMI and ESD applications we prefer the higher frequency ferrite material (Fair-Rite Type 43/Steward Type 28). For lower frequencies, such as conducted interference from power supplies, we prefer the lower frequency material (Fair-Rite Type 73). Be sure to select the core material appropriate to your needs. Low frequency ferrites, such as used in power supplies, are not suitable for high frequency filtering.

Ferrite materials can be used on individual lines for DM (differential mode) filtering (attenuates any noise current on that line) or on entire cables for CM (common mode) filtering (attenuates CM currents on the cable, but does not affect DM currents.) In the latter case the ferrites are often "clamp on" devices that can be fitted over an existing cable. Since many high frequency problems are due to

Figure 4-6. Ferrite Impedance vs. Frequency

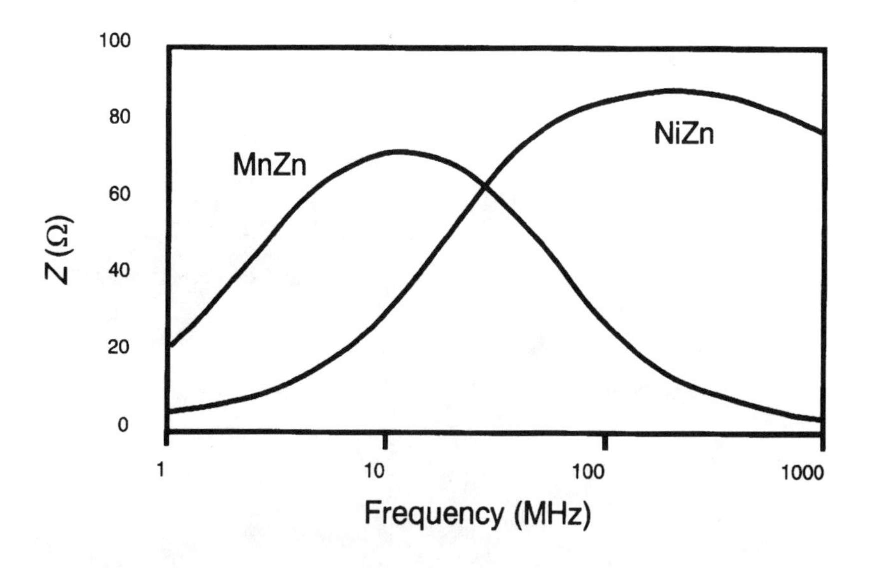

CM cable currents, the CM ferrites are effective in a wide range of situations. Figure 4-7 shows some surface mount ferrites suitable for PCB (printed circuit board) mounting and Figure 4-8 shows some larger ferrites suitable for cable mounting, many of which are split for mounting after the cable has been assembled.

For additional ferrite attenuation, multiple turns can be used. The impedance of the inductor varies as

$$Z = KN^2 l \times \ln(OD/ID)$$

where K is a constant of the core material, N is the number of turns, l is length, OD is outer diameter, and ID is inner diameter.

This works for both DM and CM ferrites. The advantage is that the inductance/resistance increases rapidly. The disadvantage is that the interwinding capacitance also increases with these multiturn windings, which severely limits the upper frequency range. As a result, we usually do not recommend more than two or three turns through a core, and then only when the major problems are below 100 MHz.

Another ferrite limitation is saturation. Since ferrites are high permeability devices, they will saturate at high current levels. This

Figure 4-7. Surface Mount Ferrites (Photo courtesy of D.M. Steward Company)

Figure 4-8. Cable Mount Ferrites (Photo courtesy of Fair-Rite Corporation)

can be a serious problem when ferrites are used to suppress DM currents. In this case the flux from any high level DC or AC (signal or power) can saturate the ferrite, rendering them less effective. This saturation takes place from the "inside-out," as it is a function of the magnetic path distance around the wire carrying the current.

As a rule of thumb for the popular Type 28/43 materials, you can predict the "saturation radius" with the following formula: $I_{sat} = 10R$ where I_{sat} is the saturation current in amps and R is the ferrite radius in cm. This gives you a quick guideline to how "fat" your ferrite should be for a given current. For example, small ferrites about the size of 1/4-watt resistors are good to about 100 mA, while ferrites about the size of a 2-watt resistor can handle a couple of amps. Do not be overly concerned about saturation, as the resistance does not decrease to zero. Instead, it tends to flatten out at about 1/4 of the unsaturated levels. Even in saturation ferrites are forgiving.

Saturation is generally not a concern when ferrites are used as CM chokes. The ferrite encloses the signal/power current going in both directions, so the high current flux levels cancel. As mentioned earlier, DM "noise" currents are not blocked, as their flux also cancels, and the current passes unimpeded. Nevertheless, many high frequency problems are primarily due to CM currents on cables, so CM ferrites are a good cure for many high frequency EMI problems. Some problems, however, will still require DM solutions. Rarely is a problem solely CM or DM, although one will usually dominate.

One more caveat about ferrites. Since they are relatively "low impedance" devices (typically in the 50–500 ohms range), they work best in conjunction with low impedance circuits. For high impedance circuits you may need to provide shunt capacitors to provide the needed low impedance at high frequencies. More on this will be discussed later in this chapter, when we look at EMI filters.

We have seen that wire wound inductors work best below 50 MHz, and ferrites work best above 50 MHz. Large wire wound inductors will not even perform as high as 50 MHz, due to capacitance between the windings. We have also seen that ferrites are very popular components for controlling EMI and ESD currents, as they behave as lossy inductors that actually absorb unwanted EMI/ESD energy. We have also seen that we must be careful not to saturate ferrites, yet we must provide a low impedance termination at high frequencies.

Resistors

Resistors are probably the most widely used of components, and yet they also have some serious EMI limitations. There are three types of resistor construction: wire wound, carbon composition, and film. As expected, wire wound resistors are not suitable for high frequency applications, due to their inherent inductance. Film resistors also exhibit some inductance, although at values under 2 kΩ, they are usually acceptable for high frequency work, since their inductance is quite low.

Another EMI concern of resistors is their ability to handle electrical overstress. Thin film resistors do not handle overstress (such as ESD) very well, so they should be avoided where overstress might occur. For example, do not use film resistors in I/O lines to limit ESD currents. Rather, use carbon composition resistors, or even ferrite beads, for this type of application. We have seen cases where film resistors arced and were eventually destroyed under ESD conditions.

Like inductors and ferrites, resistors also exhibit end-to-end capacitance. This can cause high frequency leakage around the resistor, which will increase with both resistance and frequency. At frequencies above about 300 MHz, you may need to install shields between the ends of the resistors if it is necessary to maintain high impedances.

Transformers

Transformers are widely used in power systems to convert voltages. They are also used to provide EMI isolation, in both signal interfaces and power interfaces. Some transformers also include internal shields to provide additional isolation against capacitive coupling between windings.

Many medical devices also rely on power transformers to provide patient isolation against 60 Hz leakage currents. In these cases we must be very careful not to compromise that isolation when seeking solutions to EMI problems. If such a compromising situation occurs, the patient isolation requirements must always take precedence.

By their very nature transformers provide CM isolation. They depend on a DM voltage across their input to magnetically link energy to secondary windings. Thus, any CM voltage across primary windings is automatically rejected. Unfortunately, transformers are not perfect in this regard, as capacitance from primary to secondary windings does provide a path for CM coupling. As a result, transformer isolation degrades as frequency increases. This degradation can allow spikes (EFT, lightning transients) and high frequency RF energy to pass right through the transformer.

To combat this high frequency leakage, many transformers incorporate internal "Faraday shields" between the windings. By properly grounding these internal shields, high frequency noise can be rejected in both CM and DM.

The key to success with shielded transformers is knowing both where and how to ground the shields. Figure 4-9 shows several examples. In the first case there is no shield–the secondary is grounded to suppress low frequency noise. In the second case the shield is connected to the power "neutral," to provide DM isolation. In this case high frequency differential noise is intercepted and shunted to its return path. In the third case the shield is connected to earth ground to suppress CM. Double and triple shielded transformers can be used to provide both CM and DM suppression.

Incidentally, as frequencies increase to very high levels, even these internal shields begin to fail. Typically, it is difficult to get much attenuation above about 10 MHz using Faraday shields. But this is not really a problem, since filters can be easily employed at these frequencies. Furthermore, much of the power line noise and

Figure 4-9. Grounding a Transformer and Shield

Isolation Transformer

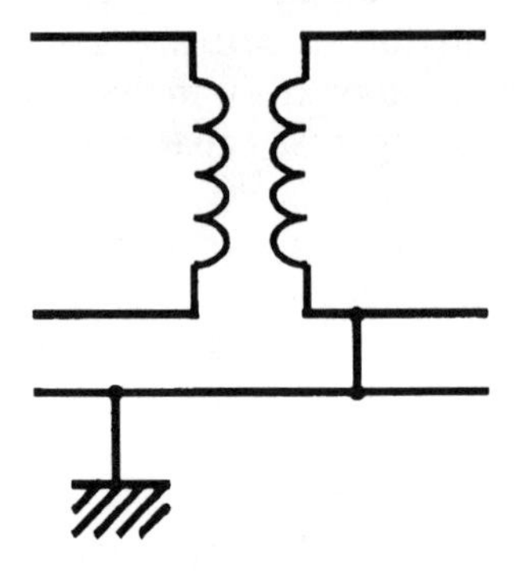

Wired to Suppress Ground/Neutral Voltages

Faraday Shield Blocks Capacitive Coupling

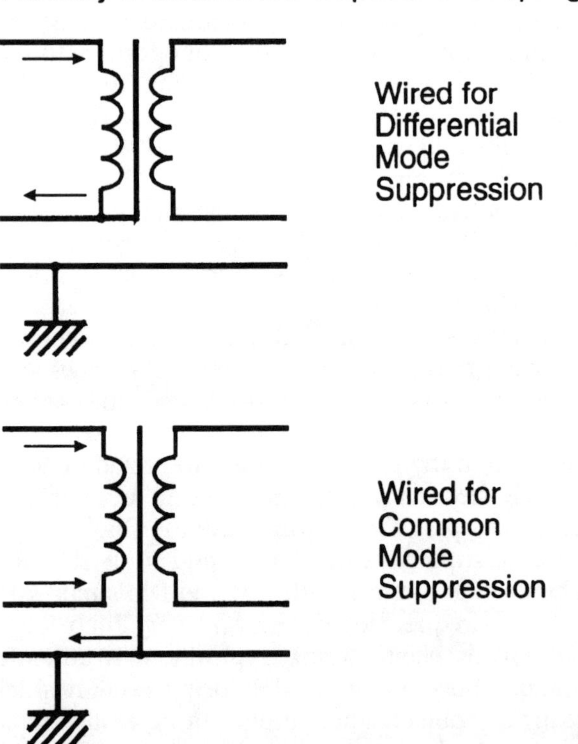

Wired for Differential Mode Suppression

Wired for Common Mode Suppression

transients are well below 10 MHz, so Faraday shields are very effective components for these types of EMI problems. In fact, we have solved many power disturbance problems with isolation transformers alone, without resorting to UPS systems or power conditioners.

The same guidelines on Faraday shields apply to signal transformers. In fact, shielded isolation transformers are popular in low level analog systems to reject 60 Hz CM noise. Even unshielded transformers can provide a lot of CM rejection at power line frequencies, and the transformers are much smaller and much more practical than filters at these low frequencies.

We have seen that although transformers degrade like other components do as frequencies increase, their rejection capabilities can be greatly increased with internal shields. The secret to success is to properly ground these internal shields. For CM rejection this should be through a short, direct connection to the chassis. For DM rejection this should be through a short, direct connection to the appropriate neutral. Finally, for power transformers in patient-connected medical devices, the leakage current limitations must not be violated.

TRANSIENT DEVICES

Many EMI problems are caused by transients, such as ESD or power line spikes. In these cases transient protection devices are often used at the signal or power interfaces to prevent these transients from entering the equipment. Like the preceding components, these special components also have limitations that must be considered if they are to be successfully employed.

There are three types of transient protection components that are popular in EMI applications. These are arc gaps, metal oxide varistors (MOVs), and silicon devices. Each of these have their advantages and disadvantages that must be considered.

Arc gap devices are the slowest, but most robust, of the transient protectors. After an arc is initiated in an arc device, the voltage drops to a very low level. This results in an almost "short circuit" condition across the terminals to be protected. As a result, most of the energy is not dissipated in the device, but is reflected back to the source. This means that a relatively small device can handle some very large transients with high energy content, and still survive. Unfortunately, arc devices typically take 10s or 100s of nanoseconds to act, which means they are generally too slow for ESD pulses in the 1–3 ns range. They are ideal, however, for slower high energy transients like lightning, with rise times in the microsecond range. In fact, arc gap devices are widely used on AC power lines for lightning protection.

Metal oxide varistors (MOVs) are faster than arc gaps, but slower than silicon devices. They are quite robust and are relatively inexpensive. Unlike arc devices that decrease in voltage, MOVs clamp transients to a fixed voltage level. As a result, they dissipate the energy in the device, rather than reflecting it back to the source. Unfortunately, most MOV devices are still too slow for ESD pulses in the 1–3 ns range, so they are used primarily for power line transients. Recently, MOV manufacturers have introduced some multilayer MOV devices for use on signal lines that work much faster, and they appear to work quite well for ESD transients–they limit voltages to something less than 100 V. Their multilayer design not only increases the speed of their operation, but also provides some shunt capacitance that also helps filter ESD transients.

MOVs have another drawback–they eventually "wear out." Each time a transient event occurs, some of the internal metal oxide modules fuse together, and are sacrificed to stop the transient. This is not a problem with most equipment, but MOVs should not be used like diodes to commutate AC, for example.

Silicon devices are the fastest devices, and like the MOVs, they clamp the transients at a fixed level. Thus, they also must dissipate the energy internally. Due to their extremely fast operation, silicon devices are well suited for even the fastest types of transients. Some manufacturers claim operation in the picosecond ranges, but these speeds are probably limited by lead length inductances. Large junction devices, such as the Transzorb® are particularly well suited for transient protection.

In all cases it is important to limit the lead length. Like capacitors, long lead lengths will minimize the effectiveness of even the fastest transient protector. Figure 4-10 shows how to limit transient lead inductance when dealing with fast transients like ESD.

ACTIVE COMPONENTS

In this section we will look at some EMI characteristics of active devices. Like passive devices, we will see that the EMI problems are very dependent on frequency. In fact, many of the problems are due to physical properties we have already investigated, such as parasitic inductance and capacitance.

Some general comments are in order. First, high speed circuits generate more noise, and are more susceptible to noise. Second, fast edge rates (dI/dt and dE/dt) easily couple energy to adjacent lines or

Figure 4-10. Minimizing Lead Length in Transient Protectors

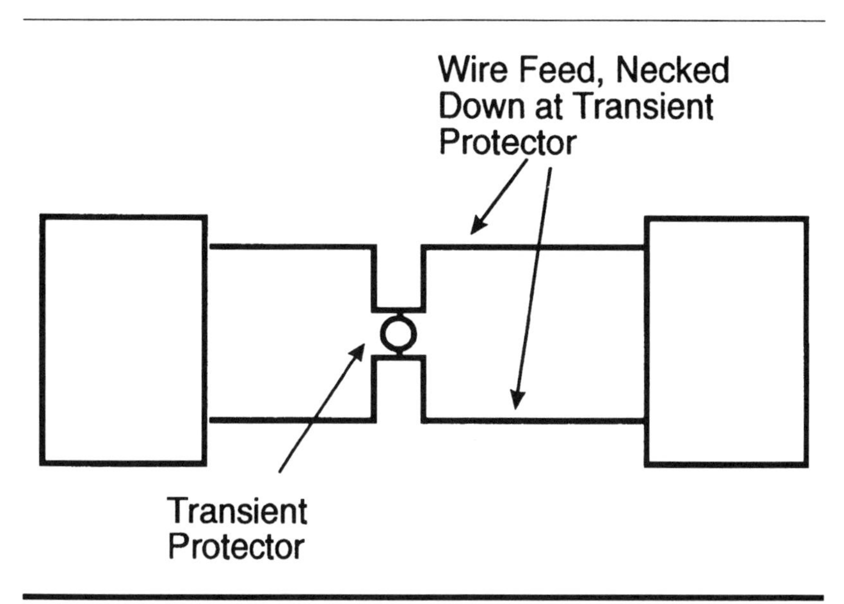

cables. Higher speeds and edge rates mean higher *bandwidth,* which means more noise gets in and out of the circuit. Today's fast circuits respond to fast transients, such as ESD or EFT. Many times our EMI efforts are aimed at limiting the bandwidth at the chip—in a sense degrading its speed, making it more like the slow chips of a few years ago.

Figure 4-11 summarizes digital failure modes of EMI. *Emissions* are a big problem with digital circuits, due to high clock and edge rates. Today's high speed microprocessor systems are notorious generators of high frequency emissions that can jam nearby radio and television communications. *Electrostatic discharge* is also a big problem, due to the fast response of most digital systems today. Damage is likely at unprotected I/O ports, and upset is common with resets and internal control lines (such as false memory read/writes) due to signal pickup and ground bounce.

Power disturbances are also a big problem. Damage can occur at the power interface due to transients, and upsets can occur due to high speed transients like the EFT, which couples through unprotected power supplies, and behaves like ESD on internal circuits. Another power effect is loss of energy due to long-term voltage sags. *Self-compatibility* is another EMI problem with digital circuits, due to crosstalk between high speed circuits, or jamming of analog

Figure 4-11. EMI Failure Modes in Digital Circuits

Digital Circuit Characteristics

High bandwidth (100 MHz+)
Moderate level (0.5 to 3 v)
Out-of-band response (1- to 100x bandwidth)
Moderate to low input Z (Ω-kΩ)

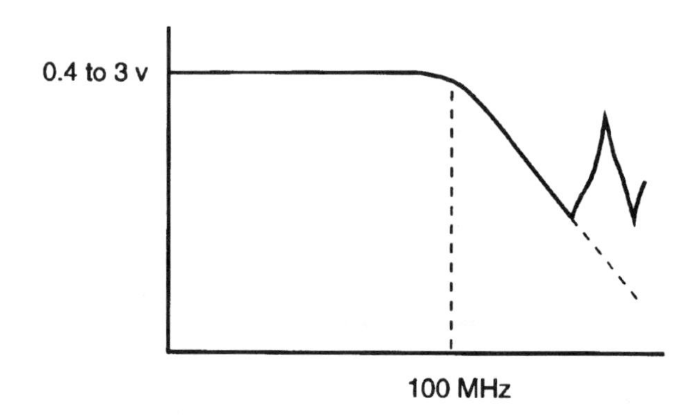

Problem Areas

Emissions-Big problem, due primarily to clocks and highly repetitive signals
ESD-Big problem, both upset and damage
RF-Upset possible if signal quite strong
Power-Upset likely due to spikes or large sags
Self Compatibility-Digital to digital through crosstalk, spikes from other noisy circuits

circuits by digital noise. Although *RFI* from nearby radio transmitters can upset digital circuits, they are not nearly as sensitive as analog circuits, which we will look at shortly.

Most of the EMI problems associated with digital circuits are due to their inherent high bandwidth. With rise times in the 1–3 ns range, this is equivalent to bandwidths in the 100–300 MHz range. As rise times continue to shorten these bandwidths will continue to increase. At 300 ps, the bandwidth is 1 GHz. Soon, many of our digital EMI problems will be microwave problems.

Fortunately, most digital circuits have moderate noise margins, so they can tolerate a fair amount of noise. Typical noise margins are in the 400 mV(worst case) to 1 V range. The move to 3 V logic, however, will reduce these margins, so we expect to see increased EMI immunity problems with 3 V logic families. Fortunately, the emissions should be lower, since the total edge rates (*dI/dt* and *dE/dt*) will be reduced by 40 percent. This will translate to about a 4–5 dB reduction, assuming the rise times are the same. It is not a lot, but it will help.

Figure 4-12 summarizes analog failure modes of EMI. We are assuming low level, low frequency analog circuits, such as those used in physiological monitoring like ECG, respiration, EEG, and so on. Many medical devices incorporate these types of sensitive analog circuits. High speed analog circuits, such as computer video circuits, behave more like high speed digital circuits, and should be treated as such for EMI purposes.

Emissions are usually not a problem with analog devices, due to their low operating speeds. For the same reason, *ESD* upset is usually not a problem, although damage can occur to unprotected analog I/O circuits. *Power* transients are not a major problem either, although analog circuits can be upset by long-term sags and surges that can "modulate" sensitive analog circuits. We have seen this problem several times, but the cure is simple–provide a local regulator for the analog circuitry.

RF Interference from nearby transmitters is a big problem with analog circuits. In fact, RFI has been blamed for a number of medical device malfunctions, ranging from patient monitors to wheelchairs. The problem is that high levels of RF energy can drive unprotected low level analog circuits into nonlinear operation. The problem is even more insidious with modulated RF, as the modulation is then "detected" and passed along to the next stage as a "legitimate" signal. For example, RF signals that are modulated can fool monitors by providing false signals in the analog processing chain.

Incidentally, this RFI problem is not unique to medical devices. It is a very common problem with industrial controls, vehicular electronics, and other electronics that contain low level analog circuits. The secret is to keep RF energy out of analog circuits in the first place. Furthermore, when designing analog circuits, one must design for high frequency threats, not just low frequency operation. Many interference specifications, both medical and nonmedical, now require that equipment be tested for RF immunity.

Self-compatibility is another problem area with analog circuits. Although they are not typical sources, analog circuits are victims of nearby noise from digital circuits, switching power supplies, and

Figure 4-12. EMI Failure Modes in Analog Circuits

Analog Circuit Characteristics

Low bandwidth (Hz-kHz)
Low level (μv-mv)
Out-of-band response (1000 to 100,000x bandwidth)
High input Z (kΩ-MΩ)

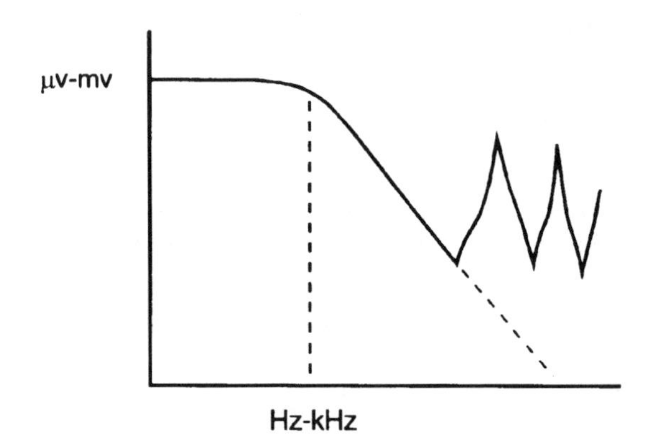

Problem Areas

Emissions-Usually not a problem
ESD-Upset not likely, but damage possible
RF-Big problem due to rectification
Power-Upset likely due to slow sags and surges
Self Compatibility-Upset likely due to digital or other noisy circuits on same power/ground

often 60 Hz power line currents and fields. The lower the analog signal level, the more sensitive the problem. Thus, low level analog circuits may fail, while higher level analog circuits may work just fine. It is a matter of "signal-to-interference" ratio. Normally, we like to keep the interference levels below the lowest level at which the circuit must operate. Thus, if we are looking for 100 μV signals, the "noise" must be kept well below this level. On the other hand, if we are working with 100 mV levels, we can withstand a lot more interference before operation is degraded.

Remember that most analog circuits typically operate at low levels and low frequencies, and yet have very high input impedances. They are also vulnerable to "out-of-band" responses, which typically occur at frequencies 1000 to 100,000 times their normal operating frequencies. These are due to parasitic resonances, and also contribute to RF interference problems. Treat your analog circuits with care, and do not assume they just work at low frequencies. In today's complex electromagnetic environment analog circuits must be able to withstand some pretty high frequencies, too.

EMI FILTER DESIGN

In the final section of this chapter, we will look at some simple EMI filter design concepts. Incidentally, for most EMI applications, you can forget about sophisticated design techniques (like Butterworth, Tchechebev, etc.) and rely on more simple and intuitive approaches. Besides, the sophisticated approaches assume known and controlled terminating impedances, a luxury we seldom if ever have with EMI filters.

Most EMI applications use a simple low pass filter. This consists of one or more high frequency shunt elements (capacitors), and one or more high frequency series elements (inductors, resistors, or ferrites.) It is imperative that these elements be selected and placed to be functional at all problem frequencies. In some cases we may even need multistage filters if a wide frequency range must be covered.

The objective of an EMI filter is to provide a maximum impedance discontinuity at the node to be protected. Thus, low impedance filter elements should face high impedance source or load impedances, and high impedance filter elements should face low impedance source or load impedances. This is shown in Figure 4-13. For low source and load impedances, even a single "high impedance" element like a ferrite or inductor will be effective. Note also that the popular "bypass capacitor" is only effective in cases where both the source and load impedances are relatively high.

Most EMI filters require two or more elements to be successful. L filters are popular, and work best if the capacitor faces the higher impedance, and the series element faces the lower impedance—this is the typical case with a low impedance driver and a high impedance receiver. T filters and π filters are also popular, and are also subject to our warnings about impedance.

Note that the shunt capacitive leg must maintain a low impedance. Remember our warnings about lead length in capacitors. When making EMI signal filters, you must keep the capacitive lead length as short as possible—no exceptions! If not, you are just

Figure 4-13. T, π, and L Filter Preferences

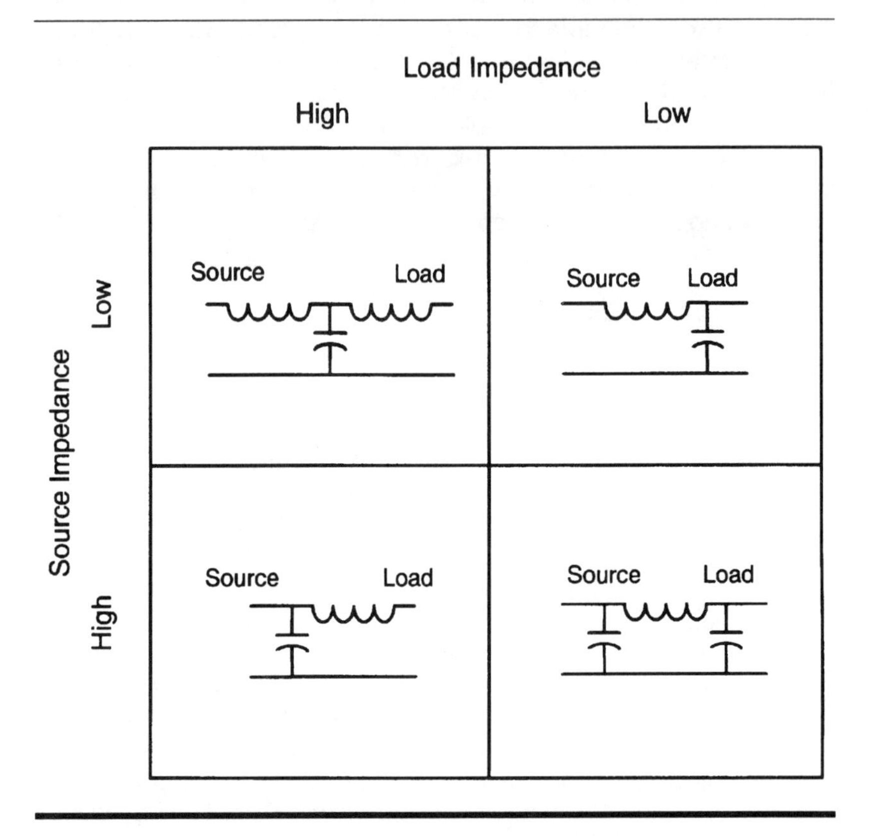

engaging in wishful thinking. We have seen dozens of problems with filters due to excessive lead length on the shunt element. Do not let it happen to you.

Any high frequency EMI filter should be placed as close to the source (for emissions) or receptor (for immunity) as practical. The objective here is to limit loop sizes, and also to limit the degrading effects of the ground impedance.

Multistage filters may be necessary. Most simple filters work over a range of about 2–3 decades before the limitations of parasitic capacitance and inductance limit their performance. To cover the entire RF frequency range of 10 kHz to 1 GHz (required in some EMI specifications), you will likely need at least three stages of filtering, as follows: 10 kHz–1 MHz, 1 MHz–100 MHz, and 100 MHz–1 GHz. We call this the "woofer-midrange-tweeter" approach. Just like audio speakers, multiple stages are needed to cover the entire frequency range of interest. Even if you are only concerned with the VHF/UHF range of 30 MHz–1000 MHz, you may need two stages of filtering. And remember, keep those capacitor leads short!

You do not need sophisticated analytical tools to design a simple EMI filter. You can obtain a pretty good approximation by using the absolute values of the impedances, and determining their mismatch at the frequency of interest. For example, a series resistor of 150 ohms and a 1 nF shunt capacitor should provide a mismatch of 150:1.6 at 100 MHz, or almost 40 dB of attenuation. A 100-ohms ferrite at 100 MHz would work about the same, and would have the advantage of not affecting the lower frequency performance due to the series resistance. This quick and dirty design approach really works—we use it regularly to solve EMI problems.

SUMMARY

Interference ultimately ends at the chip: Circuits will cheerfully amplify any voltage between the signal and ground pins. Much of EMI control involves minimizing unwanted energy at or near the chip, notably filtering. Let us look at the key points at the circuit level:

- Components degrade at high frequencies.
  - Wires become inductive then antennas.
  - Capacitors become inductive.
  - Inductors become capacitive.
  - Resistors become capacitive or inductive at various frequencies.
- Select high frequency capacitors (usually ceramics) for EMI control. Keep lead lengths short.
- Select high frequency inductors (usually ferrites) for EMI control. Be sure the ferrite material is suitable for the desired frequency range.
- Carbon or film resistors have good frequency characteristics up to about 300 MHz. Thin film resistors are limited in the overstress they can handle.
- Select a T, π, or L filter to achieve maximum impedance discontinuity in the circuit. Use series resistor or ferrite and shunt capacitors.

- Mount filter capacitors to divert adverse currents away from sensitive loads and back to the source.
- Digital circuits are most vulnerable to transients such as ESD and EFT. High speed digital circuits are more vulnerable.
- Analog circuits are most vulnerable to RFI and power sags/surges.
- Decouple all voltage supplies to chips, including analog devices. Use all power and ground pins on chip.

5

PRINTED CIRCUIT BOARD DESIGN

Good PCB design is at the heart of good EMI design. All EMI problems ultimately begin and end at a circuit, so careful attention to circuit board design can reap big dividends in the battle against EMI. Nowhere else is EMI control as effective. Most PCB solutions are inexpensive, and some are even free. Or as we like to say, "an ounce of prevention is worth a pound of shielding." This is especially true at the PCB level.

There are two approaches when dealing with EMI at the board level. You can practice source suppression, circuit hardening, or both. The approach will depend on your needs, so it pays to determine your EMI objectives before you begin board design. If you wait until the board is done before you begin to think about EMI, you are probably already in EMI trouble.

In this chapter we will look at circuit board design techniques to prevent many common EMI problems. We will build on the information learned in chapter 4 on components. We will examine how to layout these components to minimize EMI, and then look at different PCB construction techniques and methods. If you follow our guidelines, you will prevent most board-related EMI problems, and maybe some mysterious "operational" problems as well.

SIGNAL INTEGRITY VS. EMI

The term *signal integrity* has become very popular in the past few years. Generally applied to high speed circuits, it evokes a certain mystique. Signal integrity addresses issues new to many digital designers, like signal overshoot and undershoot, crosstalk, reflections, and so on. Yet may of these are really old-fashioned EMI issues dressed up with a new title. While these issues primarily affect the self-compatibility of a system, they can also affect EMI emissions and immunity.

A key difference between signal integrity and EMI is amplitude. Where signal integrity deals with ***millivolts*** and ***milliamps***, EMI issues often deal with ***microvolts*** and ***microamps***. Thus, designing for signal integrity may still be woefully inadequate when it comes to many EMI problems. For our purposes we will look at both areas.

EMI ISSUES IN PRINTED CIRCUIT BOARDS

In the previous chapter we saw that analog and digital devices have different EMI problems. As a result, we also need to emphasize different design techniques for analog and digital PCBs. One set of rules will not suffice. In fact, this is where many EMI problems begin, when the "wrong" rules are applied without thinking about the design objectives. First, we will look at some fundamental EMI issues for both analog and digital circuits, and then look at how to prevent and solve these problems during the PCB design stage.

Many medical devices today include both analog and digital circuits. Analog circuits are often used for detecting and conditioning very low level biological electrical signals, such as EEG, ECG, or respiration signals. Although low in level (in the microvolt to millivolt range), these waveforms are relatively slow, with bandwidths below 1 kHz. On the other hand, digital circuits, such as microprocessors, have very high bandwidths, and operate at higher levels. As we have already seen, both are vulnerable to outside EMI sources such as ESD, RF, or power disturbances. Digital circuits, however, are also significant sources of EMI. They radiate, causing interference to nearby communications equipment, and they also try to jam their analog neighbors.

As a result, analog and digital PBCs demand different EMI emphasis. And since the two must eventually communicate with each other, special attention is often needed at the analog/digital interface.

DIGITAL CIRCUITS

As we saw in the previous chapter, high speed digital circuits are rich sources of emissions, and are vulnerable to "spiky" external disturbances, such as ESD, power transients, and even crosstalk from adjacent digital circuits. Digital circuits have an inherently high bandwidth, reaching into the hundreds of megahertz at today's nanosecond speeds. (Remember EFFT—a 1 ns edge rate is equivalent to just over 300 MHz.)

As a result, high frequency design techniques must be applied to digital circuit boards. Traces often act as unwanted "antennas" that both transmit and receive electromagnetic radiation. Furthermore, trace inductance results in high impedances at high frequencies, causing power and grounding problems. Thus, we often need to use planes, rather than traces, for power and ground distribution. We also need to control "loop" sizes on traces to minimize antenna effects.

Figure 5-1 shows some typical radiation and coupling paths on PBCs. Note that there are *multiple* paths, not just one. Too often, designers waste time looking for the "magic bullet" fix, when many fixes are needed. The same is true of preventative design; you need to consider all the possible leakage paths. It is a bit like fixing a leaky boat. Until you plug all the leaks, you will never stay dry.

The most logical area of concern is radiation from *signal traces*, or loops. The signal currents flow from sources to loads, and then return to their respective sources. These paths behave like small "loop antennas," and will support radiation. They will also act as antennas to receive electromagnetic field energy, such as that from radio transmitters or a nearby ESD event. Thus, controlling these loops is paramount to controlling EMI emissions and immunity at the PCB.

A related area of concern is radiation from *power traces*, or loops. Although less obvious than signal trace loops, power traces often cause much higher emissions than signal traces. Most people

Figure 5-1. Coupling Paths on PCBs

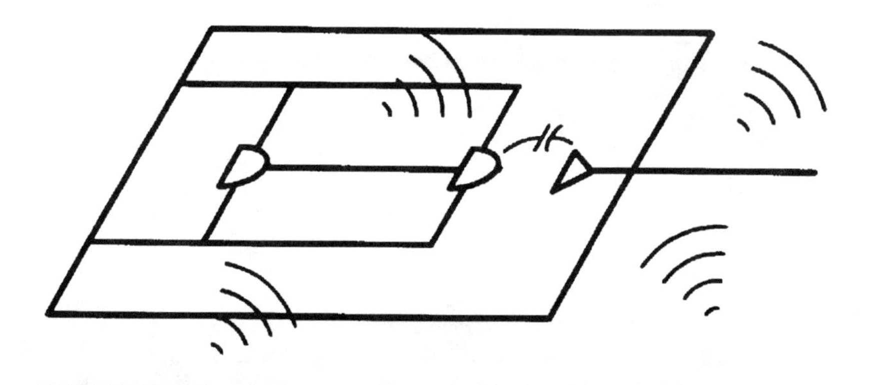

incorrectly assume that power traces are carrying DC, but they actually are overlaid with currents that are pulsating at the switching rate of the digital device. Thus, power traces on a microprocessor are actually pulsing at the internal clock rate, and consequently, can and do radiate. Local power decoupling is essential to control this critical emission problem.

In addition to direct board radiation, these loops can also couple energy to other circuits and cables, which then reradiate the energy. If energy is coupled to an I/O or power cable, this can be disastrous. First, the cable is much larger, so it is much more efficient as an antenna. Second, the cable current flows in "CM," which means there is no nearby return current to help "cancel" the radiation. The net result is that even a small fraction of the initial energy can cause problems if it gets onto a long cable.

Figure 5-2 illustrates this problem. We can predict CM and DM emissions from a circuit board with small current loops, connected to a cable. In this case we will assume just 0.1 percent of the current from one circuit is converted to CM current on the cable. Using antenna theory, we can predict the electric field levels from both the circuit board and the cable. For the DM case the "antenna" is modeled as a small loop, while for CM the cable is modeled as a short dipole. Table 5-1 shows the results of this prediction.

Figure 5-2. Comparison of Common Mode and Differential Radiation

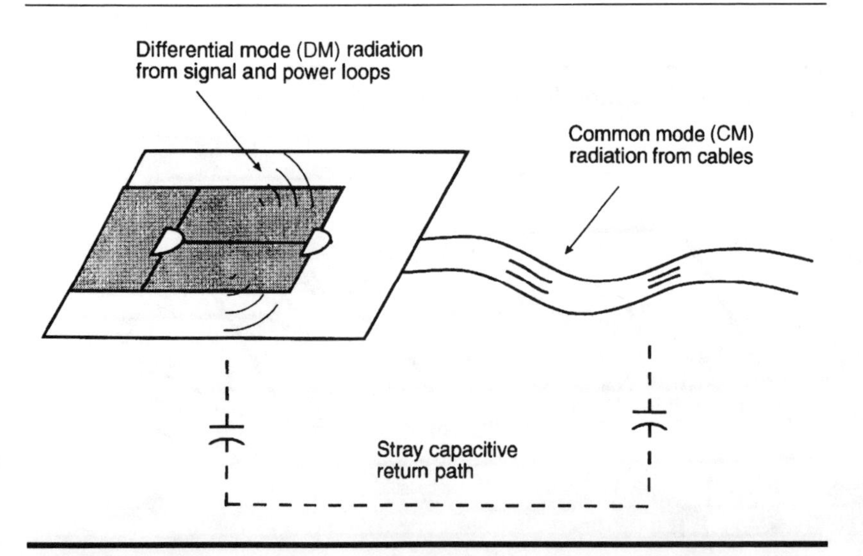

Table 5-1. CM and DM Predictions

Differential Mode	Common Mode
$\boldsymbol{E} = (265 \times 10^{-16}) iAf^2/r$	$\boldsymbol{E} = 4\pi \times 10^{-7}) ifl/r$
f = 100 MHz	f = 100 MHz
i = 25 mA	i = 25 μA
r = 3 m	r = 3 m
$A = 1\ cm^2 = 10^{-4} m^2$	l = 1 m
$\boldsymbol{E}$ = 220 μV/m	$\boldsymbol{E}$ = 1 mV/m

Even at 0.1 percent, CM radiation still exceeds DM radiation by five times. Alternately, in this example (Table 5.1) we could say that CM radiation is 5000 times stronger than the board radiation. By reciprocity we can also deduce that CM reception is also more efficient.

Thus, in addition to controlling PCB loop sizes, it is imperative that we block EMI energy at points of entry or departure from the boards as well. This is can be done with filters and other high frequency decoupling at these entry/exit points. We also need to prevent sneak coupling onto I/O circuits through common ground or power impedances, or through crosstalk between traces. We will look at some of those solutions later in this chapter.

ANALOG CIRCUITS

As mentioned in chapter 4, analog circuits are usually not a source of emissions, but they are very vulnerable to both internal and external EMI. The smaller the operating signal level, the more vulnerable the circuit to EMI. For example, if you are looking for a 10 μV signal, then you cannot tolerate 1 mV of interference. On the other hand, if you are looking for 100 mV signals, then 1 mV may not even matter. This means that we must usually pay more EMI attention to the lowest level stages. When dealing with analog circuits, always keep the "signal-to-noise" concept firmly in mind.

In addition to signal levels, we must also be concerned with frequency levels. Even though analog circuits normally operate at low frequencies, they may be exposed to high frequency threats. In fact, most analog circuits are very sensitive to high levels of RF energy, due to a failure mode known as "audio rectification." In this case high levels of RF cause nonlinear operation (the extreme case being driving an analog circuit into saturation or cutoff), which can then "detect" any low frequency modulation and pass it along to subsequent stages.

In the presence of high levels of RF energy, analog circuits often behave like an old-fashioned crystal radio. As a result, we often need to use high frequency design techniques with these normally low frequency circuits. Both the FDA and the EU specify tests for RF immunity. In our experience analog circuits fail at RF levels well below those of digital circuits.

Thus, high frequency design techniques are often mandatory for low frequency analog circuits. Just as with digital circuits, you need to pay attention to loop sizes, and you need to be very careful with external cables that connect to analog circuits, such as EEG or ECG inputs. Filtering these lines is often very limited, due to the extremely low leakage currents allowed in patient-connected devices. In short, the simple, slow analog circuits used in medical devices pose some very challenging EMI problems.

Analog circuits are also often affected by the electric and magnetic fields associated with 50/60 Hz power. At these low frequencies we can treat the coupling as if it were due to parasitic capacitance (electric fields) or parasitic mutual inductance (magnetic fields). The coupling is impedance related–in high impedance circuits, coupling is predominately capacitive, while in low impedance circuits, coupling is predominately inductive. Since most analog circuits are high impedance, coupling is usually capacitive in nature. For these low frequency fields we may need to include some low frequency capacitive shielding to minimize coupling.

WHERE ANALOG MEETS DIGITAL

Many EMI problems occur at the analog-digital junction. As mentioned above, digital circuits can easily jam low level analog circuits, so this interface is critical. You must pay attention to three areas–signal, power, and ground. If isolated circuits are used, then you must also be sure to isolate all three areas.

Figure 5-3 shows a common problem. In this case the digital return currents cause a voltage drop across the ground impedance shared by both the analog and digital circuits. If this voltage drop is large enough, small analog signals can be jammed. Although the analog ground path is usually the most critical (since the signals are referenced to the analog ground), similar interference can occur due to a shared power feed.

Figure 5-3. Common Impedance Coupling

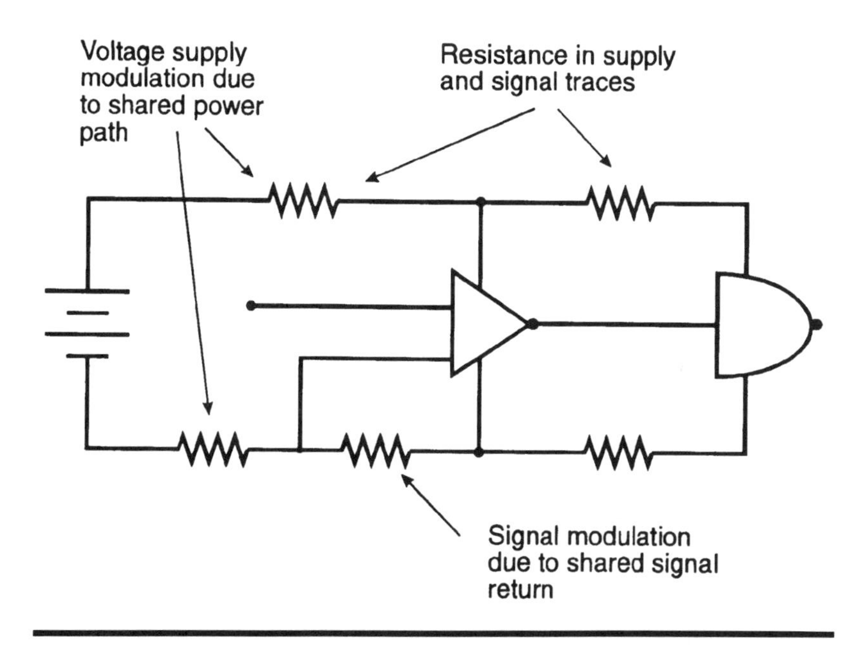

Figure 5-4 shows some solutions to this problem. In this case separate power supplies are provided for the analog and digital circuits. Furthermore, only one common ground point is used between the two circuits. By using one, and only one, common ground point, digital currents are prevented from flowing in the analog ground.

A word of caution: If someone decides to connect the two power supply grounds together (at point X), then you no longer have a single-point ground connection. In this case digital ground currents can once again flow through the analog ground. (This is the dreaded "ground loop" that we will meet again in chapter 6 on grounding.) At higher frequencies even parasitic capacitance between the power supply wiring can form an alternate "high frequency" connection. Now let us move on to some general PCB design guidelines.

PCB PARTITIONING AND CRITICAL CIRCUITS

The first concern of any PCB design is where to put the circuits on the board. Although it almost sounds trivial, your EMI success will hinge on how well this first stage is implemented. Different circuits

Figure 5-4. Solutions to Common Impedance Coupling

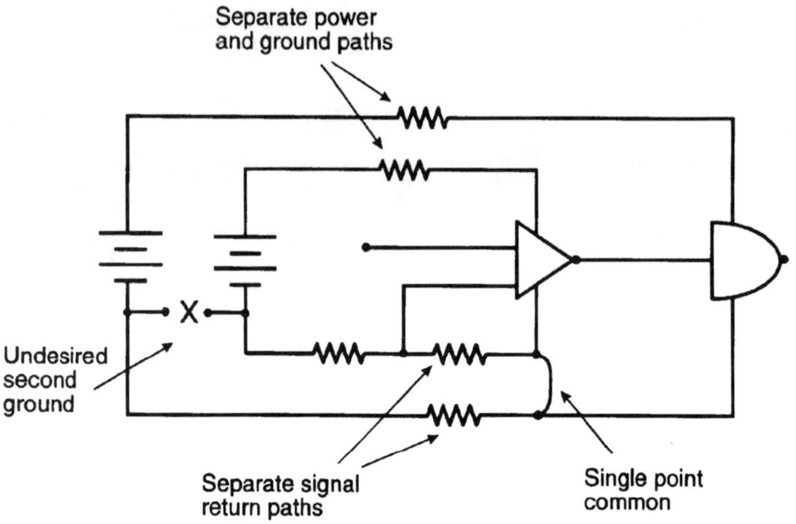

often interact in subtle and unplanned ways, so you need to plan your layout. Do not just hand the schematic to the PCB designer without proving some guidance in this area.

Circuits should be grouped together on PCBs by speed of operation. While this seems like common sense, it is amazing how often this simple concept is ignored. Figure 5.5 shows an example of a PCB partitioned by frequency and speed. Note that in this example, the high speed digital circuits are separated from the lower speed digital and analog circuits. Furthermore, the high speed circuits are separate from the I/O connectors to minimize parasitic high frequency coupling to I/O cables.

A corollary to partitioning is identifying the most critical circuits. Our experience shows that most of the problems are caused by a few key circuits. The good news is that by concentrating on these key circuits, we can prevent or solve many EMI problems with very little effort.

For *emissions* the critical circuits are almost always related to clocks, buses, or other fast, highly repetitive waveforms. These alternate circuits include ALE (address latch enable) lines, and sometime even address bit 0 on memories. All of these signals are rich in high frequency harmonics and are the most common source of high frequency emissions.

Figure 5-5. PCB Partitioned According to Speed

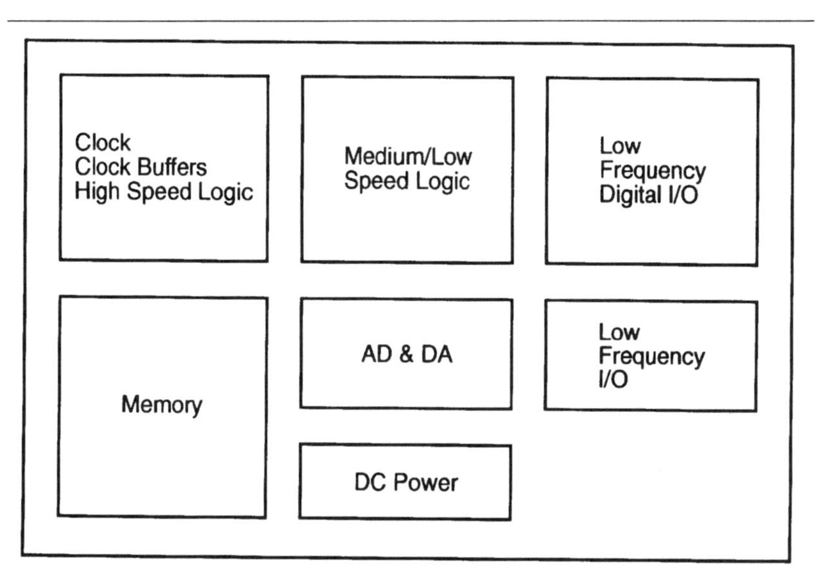

For *immunity* the critical circuits are typically resets, interrupts, or control lines which go off board. Even though these circuits may have slow (or even nonexistent) repetition rates, they are still very vulnerable to "spiky" interference that results in false triggering. Fortunately, since these circuits do not typically need to operate at high speeds, simple high frequency filtering of these lines can solve these problems. For example, we have solved many irritating ESD problems with a series ferrite (or even a 100-ohm resistor) and a 1 nF shunt capacitor on these lines.

Remember, good layout is key to good EMI control. Separate the components according to frequency; do not mix analog and digital circuits at all; and separate high and low speed lines and circuits. Route the high speed and sensitive lines first, and keep these lines short and direct. You should preferably hand route these lines, but if you use an auto-router, be sure to inspect for good routing of the critical lines. (As we are fond of saying, *auto-routers will route to maximize EMI.*)

Since clock lines are critical, position the chips to minimize clock runs. Avoid placing clock traces or oscillators near I/O ports at all cost–anything less than 25 mm and you are asking for trouble. You may need to terminate clock lines to minimize ringing, which we will look at later in this chapter.

Some of the same advice applies to resets, interrupts, and control lines. Position these to minimize crosstalk, and avoid placing them near I/O ports to minimize pickup from external EMI. Use high frequency filters to slow down the response.

PCB CONSTRUCTION

The next critical decision (after deciding on where to locate components) is how to build the circuit board. Key elements are multilayer vs. two layer, how to route traces, how to decouple power, and whether to isolate planes or sections of the board.

Before we begin this section, here is a brief editorial comment. Many high frequency EMI problems can be traced to using two-layer boards. With the new FDA and EU emphasis on EMI immunity, ***we strongly recommend multilayer boards for medical device designs, for both analog and digital circuits.*** Multilayer boards will minimize, and often eliminate, EMI problems with RF immunity, RF emissions, and ESD.

While we realize that two-layer boards are cheaper than multilayer boards, the overall project cost may actually increase with two-layer boards. An extra round of EMI testing can easily cost $25,000–$50,000 or more for redesign and retest. Most medical device designs do not have the high volumes that justify the added risk and additional test and engineering costs of two-layer designs. Thus, we will focus here on multilayer designs, but we will include some two-layer design guidelines for those of you who insist on following that approach.

Figure 5-6 illustrates the advantages of multilayer boards over two-layer boards. First, the power and signal "loop areas" are minimized, reducing PCB–radiated emissions and susceptibility. Second, the power and ground impedance levels are lowered, often by orders of magnitude. Third, the nearby power and ground planes greatly minimize crosstalk between traces. In our experience a multilayer board provides a 20 dB (tenfold) EMI improvement over a similar two-layer board. In fact, the performance difference is so dramatic that it is difficult and often impossible to design to overcome this difference.

The multilayer miracle works because of the "image-plane" effect. This has been known for a long time by radio and antenna engineers, but has only been recently discovered by many digital engineers. Nevertheless, it worked for Marconi, and it works very well today. If you place a current carrying wire closely parallel to a conductive surface, most of the high frequency currents will return

Figure 5-6. Loop Areas in Multilayer and Two-Layer Boards

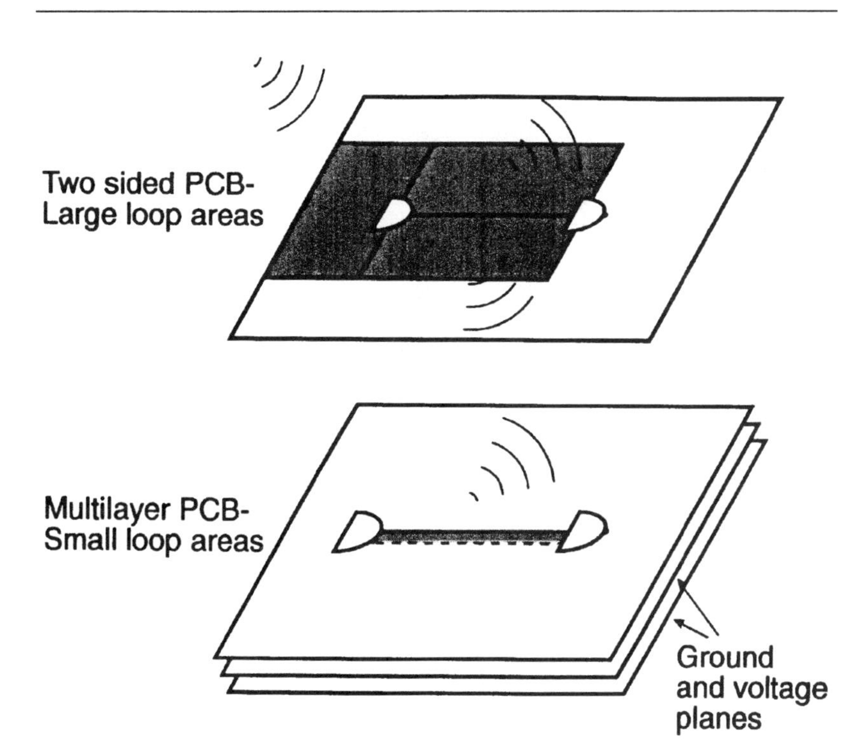

directly under the wire, flowing in the opposite direction. This forms a transmission line with the "mirror image" of the wire or trace over a plane. Since the currents are equal and opposite in this transmission line, electromagnetic radiation (emissions or pickup) is minimized.

By the way, at high frequencies both the power and ground planes act as "signal ground" planes. Since they are decoupled through bypass capacitors and their own capacitance between the planes, they are at the same potential at high frequencies.

The path under the traces (in the power or ground planes) must be continuous and unbroken. If the plane is cut, or if someone decides to "borrow" some plane area for trace routing, the return current is forced to divert the cut or break, and the benefits are quickly lost. Watch out for connector cutout areas, too, as return currents may be forced to "loop around" in this area. This is particularly insidious, since it can increase coupling to or from the cable attached to the connector. Figure 5-7 illustrates the problem in a connector area. The solution is to connect small traces across the cutout.

Incidentally, you do not need multiple planes for all of this to work. The miracle occurs with the presence of the first plane, so if you are really pressed to control costs, you can achieve many of the benefits of a multilayer board by dedicating one layer as a ground. It must be continuous, however, and if you try to sneak some traces from this plane, you will quickly lose the benefits.

Having separate power and ground planes, however, does provide further benefits. Since the power is fed from the two adjacent planes, power-related EMI is greatly reduced. The parallel planes both raise the capacitance and lower the inductance of the power distribution. In effect, the power is fed from a low impedance transmission line, with attendant EMI benefits.

Figure 5-7. Cutouts in Connector Area Create Current Loops

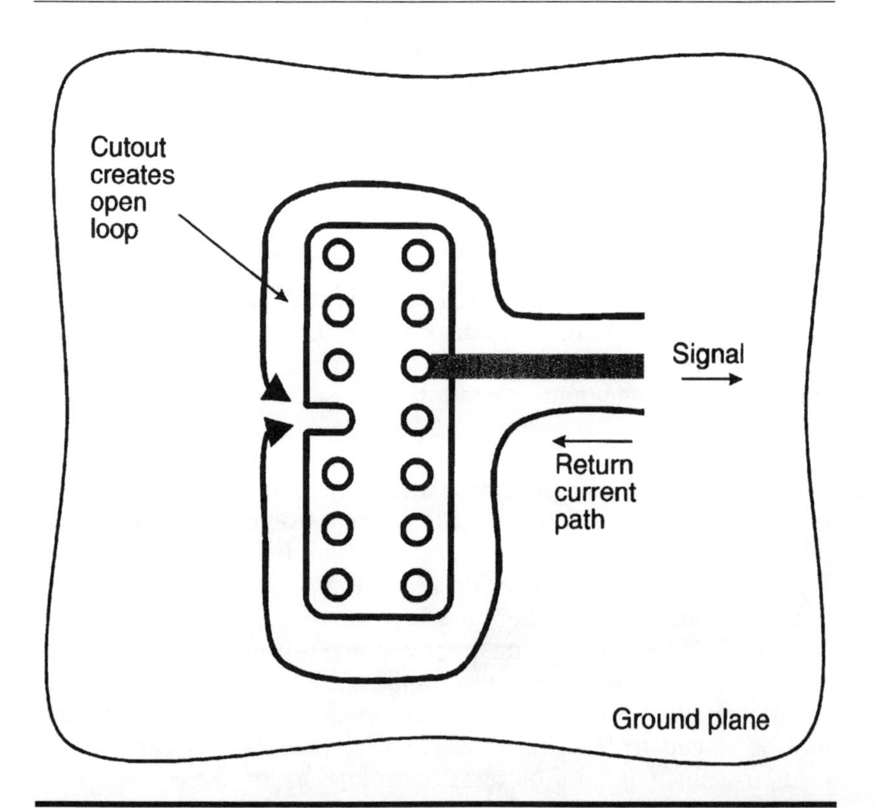

MULTILAYER BOARD STACKUP

A common question that arises with multilayer boards is how to stack the planes. For four-layer boards, should the traces be inside or outside? For more than four layers, should traces be alternated with planes or imbedded between the planes? Here are some thoughts on these issues.

For four-layer boards we generally prefer to route the signal traces outside, with the power and ground plane next to each other on the inside layers. First, this results in a lower characteristic impedance between the power and ground planes, and second, it is much easier to modify and troubleshoot the board. On the other hand, some claims have been made for reduced board emissions with the traces inside, due to shielding from the planes. In our experience, however, this shielding is minimal, and much of it is lost due to radiation from the components and their internal bonding wires. Remember that trace radiation is already greatly reduced from two-layer designs, due to the adjacent plane "image" effects.

For six-layer boards we generally prefer to stack the planes as shown in Figure 5-8. In this case we trade off keeping the traces next to an image plane, versus keeping the power and ground plane next to each other. The alternate approach of keeping the routing layers on the outside also works, but the outer layers must contain only "slow" circuits to control RF emissions and immunity. Furthermore,

Figure 5-8. Preferred Stacking on Six-Layer Boards

this preferred approach does allow one to imbed some of the more noisy or vulnerable traces, such as the clocks, resets, interrupts, and so on. Just make sure you keep the clocks away from the vulnerable traces, so you do not get some internal crosstalk.

For more than six layers once again we prefer keeping all signal traces within one layer of an associated plane for signal return currents. We still make sure that the associated power and ground (power return) planes are adjacent to each other. For example, do not put an analog power plane between digital power and ground, and so on.

ISOLATED PLANES

It has become popular to isolate planes on multilayer boards. The idea is to provide additional isolation between "noisy" and "quiet" parts of the circuit board. This technique has been used for years for boards with mixed analog and digital circuits, and we strongly recommend it for these applications. In other applications it can prove helpful, but in all cases it must be properly implemented. If you do a poor job on isolating or segmenting the planes, you may end up with more problems than if you stayed with solid planes and careful partitioning.

Two problems must be avoided. First, you must align the corresponding planes as shown in Figure 5-9. If you allow the planes to overlap, the interlayer capacitance defeats the isolation at high frequencies. Second, you must avoid crossing the boundaries with signal traces as shown in Figure 5-10. If the traces to go "over and back," you create loops that can efficiently couple noise into isolated areas. These are precisely the problems we are trying to avoid.

You must also use care in joining the isolated areas together. For low frequency signals high frequency isolation is provided if the power and ground planes are connected through a ferrite bead. The signals may also be further isolated with ferrites or common mode chokes. This does not work, however, if the signal return path must carry high frequencies. In these cases the planes should be joined together with a "bridge," with any interconnecting signal traces crossing this bridge to minimize loop areas. We generally avoid isolation with high frequency.

TWO-LAYER BOARD TECHNIQUES

Here are some recommendations for those of you who insist on using two-layer boards. If you use these techniques with care, you can

Figure 5-9. Avoid Overlapping Planes

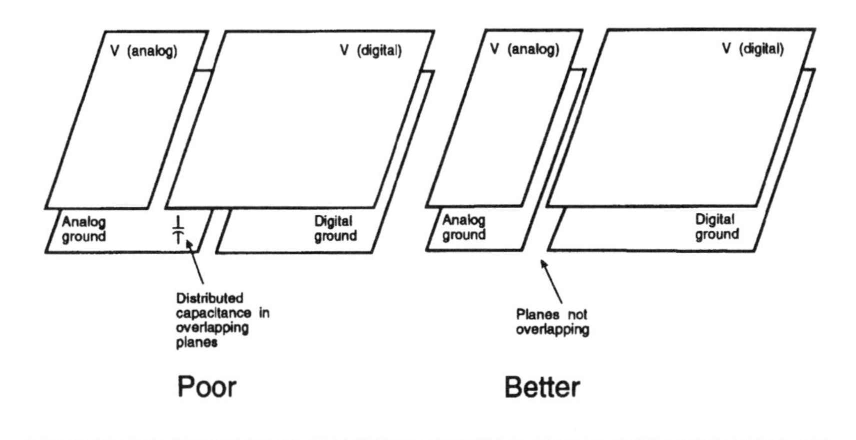

Figure 5-10. Avoid Routing Traces Over Unrelated Areas

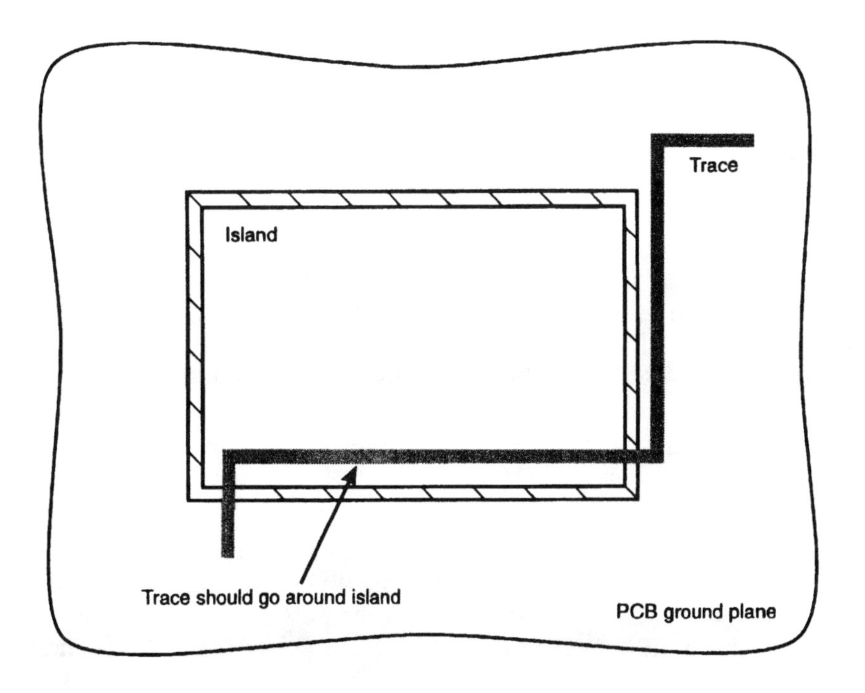

approach multilayer board performance with moderate frequency boards. We still recommend multilayer board digital circuits with clock speeds over 15 MHz or rise times under 2 ns.

First, think in terms of transmission lines. Power and return lines should be run next to each other, as they are distributed to the integrated circuits. Dedicated return traces should also be provided for critical traces, such as clocks, resets, control lines, sensitive analog lines, and so on. The objective is to minimize loop structures that radiate and pick up electromagnetic energy.

Second, use plenty of local filtering. Decouple the power on every integrated circuit, right at the circuit. Add small high frequency filtering to the critical signal lines, again right at the circuit. With two-layer boards you do not get the high frequency decoupling benefits of the parallel planes.

Creating a ground grid can also be effective in two-layer boards. In this case parallel ground traces are "east-west" on one side of the board, and "north-south" on the other side. They are connected together at the crossover points with vias to form a grid. If the grid spacing is kept below 25 mm, the grid works about as well as a plane to well over 500 MHz. By routing on both sides in alternate directions, signal trace routing problems are minimized. This technique was more popular when use of DIPs was more common.

Another technique that we have used with good success on imbedded control applications is the "micro-island," as shown in Figure 5-11. This works best if there are only a few high speed components on the board, such as a microcontroller and perhaps a RAM and ROM device. In this case one side of the board is dedicated to a ground plane (or planes if you want separate analog and digital planes). Every line (signal, power, and ground) entering or leaving the "island" is filtered through a series ferrite or resistor, and shunted with a small high frequency capacitor (typically 100–1000 pF) to the local ground plane. The high speed circuits are very effectively isolated from the rest of the circuit board.

POWER DECOUPLING

Many high frequency radiated EMI emissions are caused by poor power decoupling. While everyone seems to understand the need to filter or control high frequency signal lines, they ignore the high frequency "pulses" that occur on power rails of switched components. The front door may be locked, but the back door is left wide open!

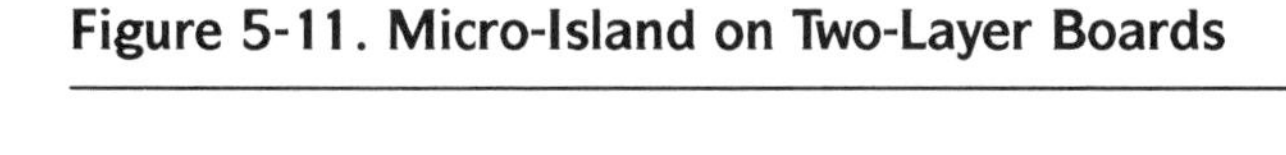

Figure 5-11. Micro-Island on Two-Layer Boards

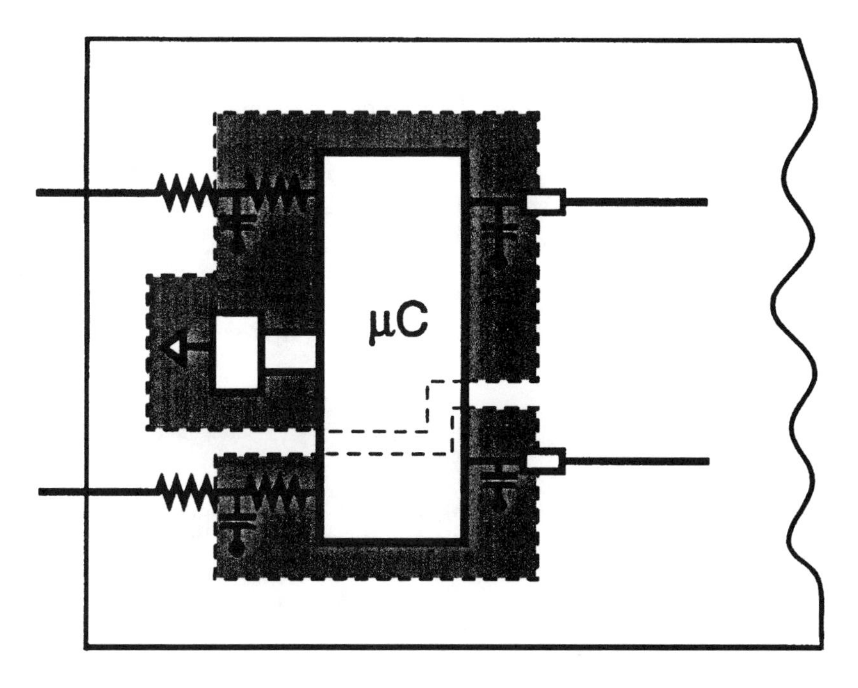

Whenever a digital circuit is switched, it "gulps" current at the switching rate. These pulses of power current radiate just as effectively as pulses of signal current. In fact, they often cause more radiation, since the power current levels are usually much higher than the levels on an individual clock line. High speed CMOS devices are particularly bothersome in this respect, since they exhibit high peak currents due to the momentary "short" that CMOS circuits present to the power supply during switching. In fact, these peak power currents can be higher than other technologies, causing emissions to actually go up when HCMOS devices are used to replace TTL technologies. This increase is in spite of lower overall power consumption. The EMI emissions are caused by peak currents, not average currents. The solution is to pay attention to the power decoupling of all CMOS devices.

We recommend local power decoupling of every integrated circuit on the PCB. For devices with multiple power and ground pins, every pair of pins should be decoupled. As a minimum, high frequency capacitors in the 10–100 nF range should be used. Extra protection can be provided with a series ferrite at the V_{CC} line as shown in Figure 5-12. This must be installed on the V_{CC} side of the capacitor, not on the IC side. Keep the capacitor leads short to avoid capacitor resonances, as we discussed in the previous chapter.

We also recommend high frequency decoupling at the power entry points. Many board designers provide 1–10 μF capacitors at the board entry to "recharge" the individual circuit capacitors. We like to add a 10–100 nF capacitor in parallel with these "bulk" capacitors as well. The bulk capacitors are generally useless at frequencies above 1 MHz due to internal inductances, so the high frequency capacitors can intercept any high frequency energy that is trying to sneak out the power input lines. It is very cheap insurance and it works.

SIGNAL TRACES

When signal traces become long relative to the rise times or frequencies they carry, they start to have "transmission line" problems. These include reflections, ringing, and crosstalk. As mentioned earlier in this chapter, these are often referred to as "signal integrity" problems, since they can distort and even destroy intended signals. These same problems can also contribute to other EMI problems as well.

A generally accepted criteria for transmission line problems in digital circuits is when the round trip transit time of a pulse is equal

Figure 5-12. Decoupling Critical ICs on Two-Layer PCBs

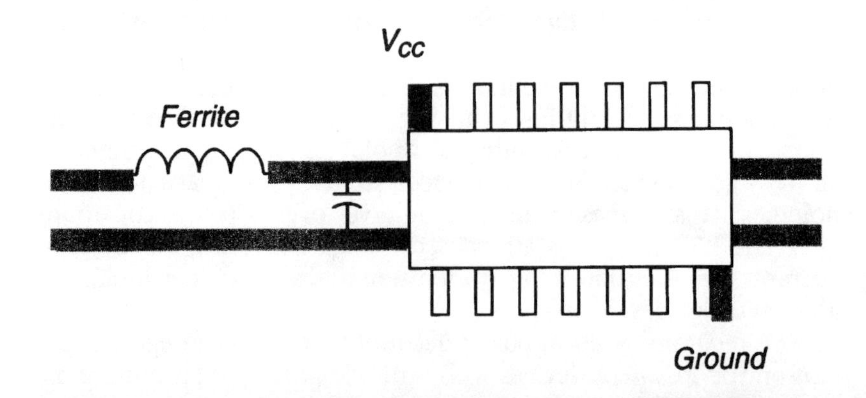

to the shortest pulse edge (rise or fall time). Any trace this length or longer will see maximum reflections due to transmission line effects. In free space the one way propagation velocity is about 30 cm/ns, which drops to about 20 cm/ns on a printed circuit board due to dielectric loading. Based on our criteria above this means that at 1 ns, any distance over 10 cm must be terminated to prevent reflections and ringing. This will also help control crosstalk, which we will explore in chapters 8 and 10 on cables and interconnections.

Figure 5-13 shows several termination techniques. The first three are aimed at matching the load impedance to the characteristic impedance to prevent any reflections from occurring. The first is not widely used, since it consumes the most power. The second, the Thevenin match, is popular, as is the AC match. With the AC match the time constant of the RC network must be at least three times the rise time if a good "short" is to be provided by the capacitor during the signal transition. All of these are popular on buses, as long as you keep any stubs short.

The fourth method, the series match, allows the pulse to reflect back to the source, where it is then terminated to prevent any further reflections. This only works for single source–single destination paths, so it can not be used on buses. This is very popular on clock lines; most designers of high speed systems routinely add a small series resistors at the clock output for this purpose. It is often referred to as a clock-damping resistor, and typical values range from 10–47 ohms. The final value is often determined experimentally by observing the clock waveform.

The fifth method shown is often used in ECL circuits. The diodes clamp the voltage to limit reflections, but they do allow current pulses to flow. Thus, this technique should not be used in traces that go off the board, since high levels of emissions may occur due to the currents.

I/O TECHNIQUES

Earlier in this chapter we saw that even small amounts of EMI currents on I/O lines can cause big EMI problems. Thus, you should provide high frequency filtering at the board periphery, particularly on those lines going to the outside world. This is similar to our advice on adding small capacitors to the power inputs, and represents the last chance to catch emissions as they leave the board, or the first chance to catch noise as it enters the board.

In recent years it has become popular to use isolated I/O ground and power planes to provide additional high frequency isolation

Figure 5-13. High Frequency Terminations

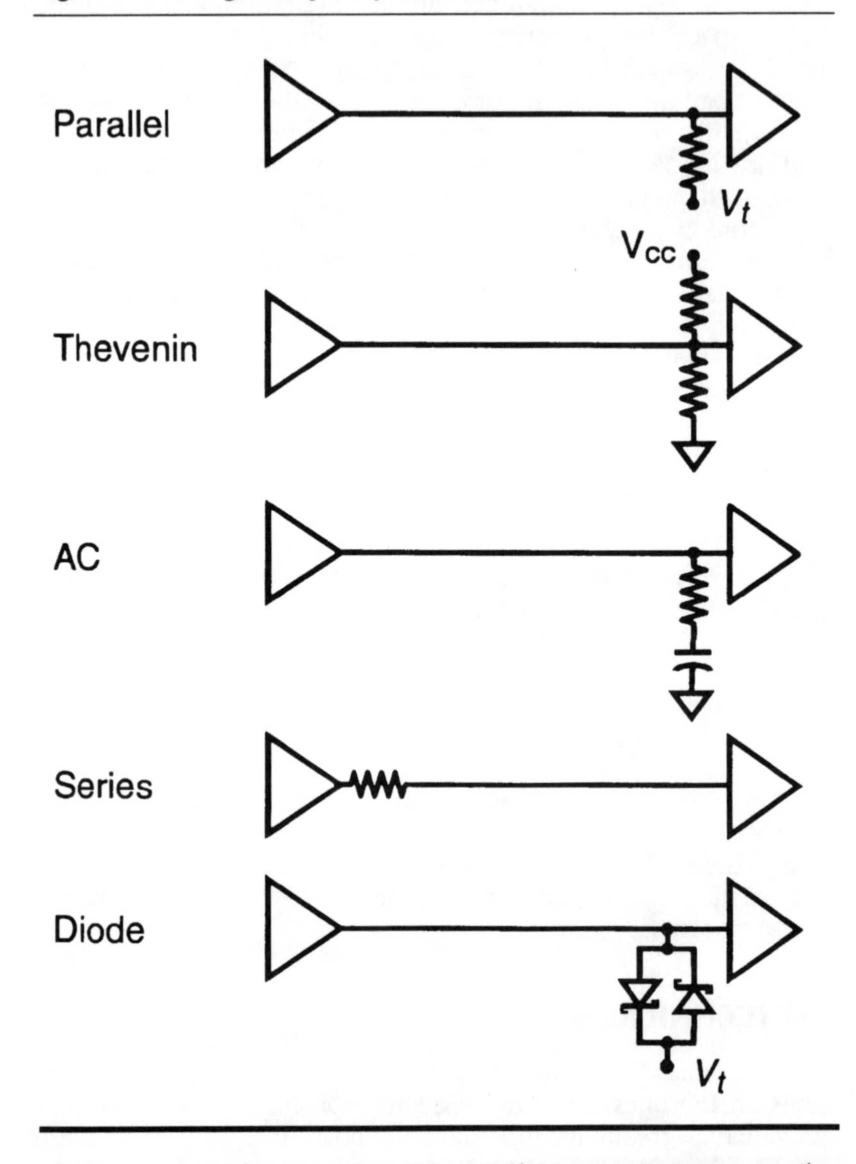

of I/O circuits. An example of this is shown in Figure 5-14. This technique is similar to the microisland technique discussed earlier in this chapter. Note that the power and ground are isolated through ferrites. In the case of ground isolation, you may need to experiment a bit here. If you have high frequency return currents (from a high

Figure 5-14. Isolated I/O Sections

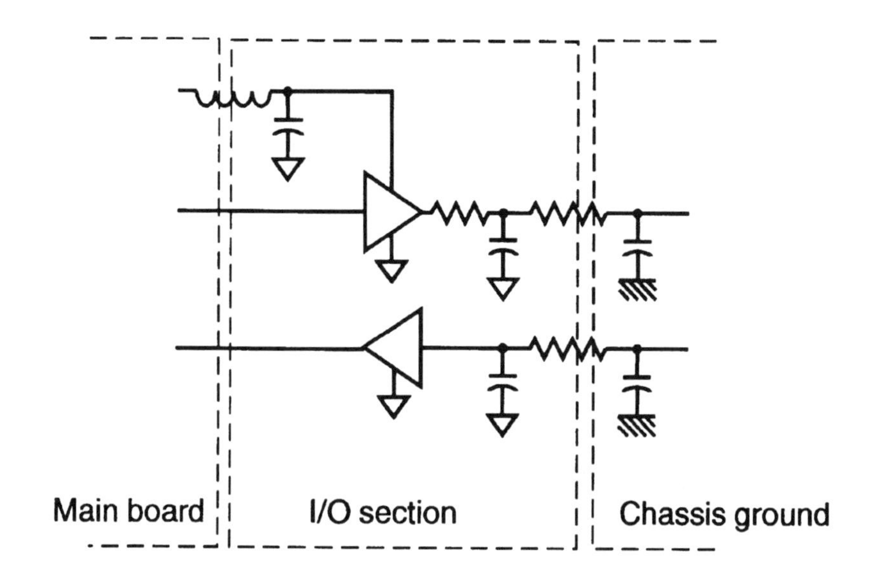

speed I/O like a video graphics interface or LAN interface), you might even make problems worse. For slow interfaces, such as analog or RS-232, however, ferrite isolation in the ground return should work well.

Here is some advice on grounding shunt capacitors on I/O filters, shown in Figure 5-15. If you are trying to contain emissions, you should connect the capacitor to the signal ground. The objective is to keep these EMI currents on the PCB. If you are trying to withstand external noise, such as RF or ESD, then you should connect the capacitor to the chassis ground, not the signal ground. The objective here is to keep the EMI currents off the PCB board.

If you do not have an external chassis, you will need to compromise. You can connect the capacitor to the signal ground, but you must provide a series element (resistor or capacitor) on the I/O side of this capacitor to limit the EMI current shunted into the signal ground. Alternately, you can provide a separate metal plate below the circuit board, and connect your capacitors to this plate. This "transient plate" approach is popular in printers and other office equipment, and is effective against ESD currents. It provides an alternate capacitive path for transient ESD currents.

Figure 5-15. Grounding I/O Filters

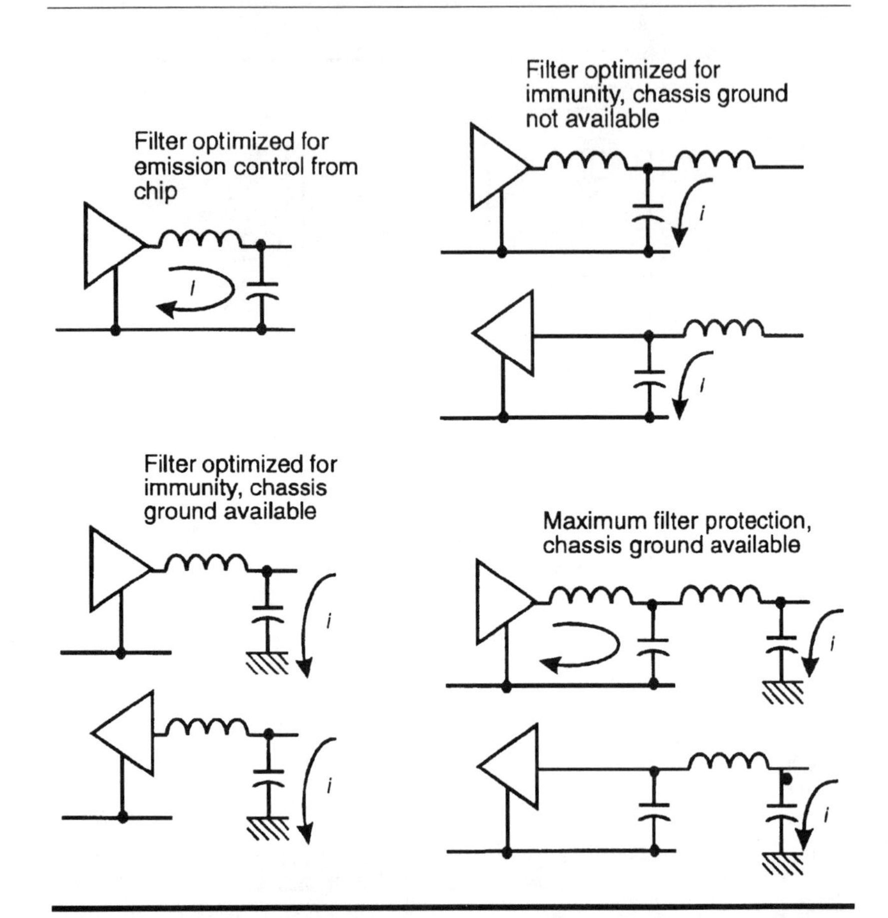

BOARD SHIELDING

Another technique that is gaining popularity is partial or even full shielding of the PCB. This can range from "fences" between critical areas, to "lids" over ICs, to fully shielded "cans" over critical circuits. Some examples are shown in Figure 5-16. Like the "image plane" concept, these techniques have been used for many years by radio designers, but they are just being discovered by high speed digital designers. Nevertheless, they are very effective, and will see increasing use as system speeds increase.

We often refer to the "can" approach as the "TV tuner" approach. The tuner is one of the most critical circuits in a television receiver. Due to low signal levels it is very susceptible to interference. It is

Figure 5-16. PCB Shielding (Photo courtesy of Instrument Specialties)

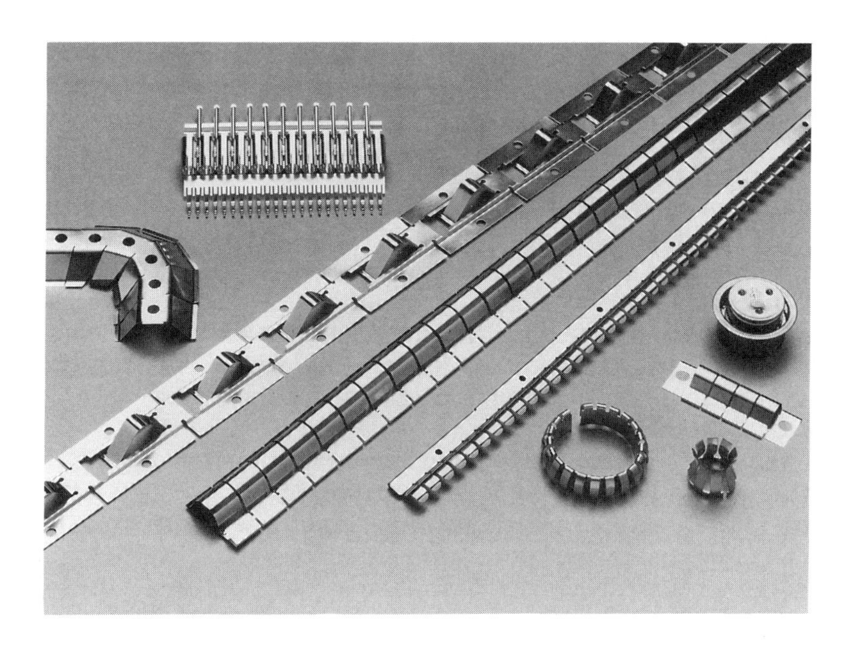

also a source of interference due to radiation from the local oscillator. Rather than shield the entire television receiver, designers for years have just enclosed the tuner in its own well-shielded box. High levels of shielding integrity are maintained by techniques like interlocking metal fingers, and filters or coaxial cables on all input and output lines.

This technique can also be applied to sensitive analog circuits in medical devices. Just enclose them in a can and filter the inputs and outputs. For patient-connected devices this metal enclosure can be electrically isolated from the rest of the chassis to maintain the required low leakage currents. This technique is useful in systems that must be patient connected, but must be subjected to high levels of RF energy.

SOFTWARE

Although not technically a PCB technique, do not overlook software as a tool in the battle against EMI. Software is normally not effective against emissions (although we have seen a few cases where it did make a difference), it can be very effective for EMI immunity. We have long been strong advocates of designing "noise tolerance" into your software. In fact, we wrote some of the first articles in this area.

The objective is to catch any errors before they upset or damage your system. After you catch an error, you should be able to recover from the error. We refer to the latter as "graceful recovery." It does not take a lot of software effort to provide this type of protection. Oftentimes, just a few extra lines of code can do wonders. If you are writing modular code, it is very easy to add noise tolerant features.

To detect program errors, you can add "tokens" to modules of code. You save a token on entering a module, and you check it on exiting a code module. If they are not the same, you have detected an error, since you entered the module illegally. To detect memory data errors, you can add "checksums" to blocks of data stored in memory. To detect I/O errors, you can do type and range checking on data.

You can add "watchdog" timers to prevent you from getting lost in an endless loop. This is often a separate device, although watchdog timers are often incorporated in "smart reset" chips that also detect low voltages and other error conditions. If the "watchdog" is not reset in a predetermined time, then it "barks" and resets the system.

Due to software safety concerns, many of these techniques are already employed in medical devices. They are not usually considered part of an overall EMI strategy, but they can be one more effective tool in the battle against EMI.

SUMMARY

Good circuit board design can be summarized with the following rules:

Board Structure

- Multilayer boards are preferred to double-sided boards (20 dB superior).
- Keep maximum ground plane area.
- Avoid large open loop areas.
- Place components to facilitate isolation of noisy or sensitive circuits.

- Segment ground/power for digital, analog, receiver, transmitter, and so on.
- Allow for plenty of I/O pins.

Bypass

- Use tantalum capacitors at headers and connectors.
- Use ceramic capacitors at headers and connectors, at each chip power pin (including analog chips), and fast transistors.
- Supplement critical circuits with ferrites.
- Reduce capacitor size if adverse resonance occurs.
- Bypass only between related power and grounds (+5 to ground, +analog to analog ground, etc.).

Routing

- Minimize trace length of clocks and other periodic signals.
- Minimize parallel runs of sensitive and noisy circuit traces.
- Keep I/O lines at edge of board, away from sensitive circuits.

Other

- Use all power and ground pins.
- Install local shields on sensitive or noisy ICs.
- Mount crystals flush with ground and ground them.
- Use differential signals wherever possible.
- Route balanced lines in parallel.
- Terminate high speed lines when needed.

Two-Sided Boards

- Use parallel runs for power and ground, with wide traces.
- Use a feed-in tree or small grid.
- Use transmission line critical signals (adjacent signal return paths).
- Place ground plane under crystal.
- Use ground fill as much as possible.

6

GROUNDING FOR EMI CONTROL

Grounding is probably the most important, yet least understood, aspect of EMI. Every circuit is ultimately connected to one or more "grounds," so every circuit is affected by grounding EMI issues. Grounding cannot be left to chance–it must be designed in from the very beginning.

Grounding issues range from individual circuits to entire buildings. This chapter will focus on the areas where the medical device designer has some control–circuit boards, circuit modules, and individual pieces of equipment.

Patient-connected medical devices impose some unique grounding constraints, due principally to power line leakage current limits. These limits are designed to prevent potentially lethal "microshocks." While these limits often pose severe constraints and make EMI even more difficult to control, safety must always take precedence.

WHAT IS A GROUND?

Before proceeding any further, some definitions are in order. One of the biggest problems with the subject of grounds is the ambiguity of the term. Ask ten different engineers for a definition, and you will likely get 20 or more answers. Furthermore, you will get as many different opinions on "how to ground" as well. One person will suggest "ground planes," while another will insist on "single-point grounds." Someone else may suggest "isolated safety grounds" or even separate ground rods in the earth. And all of them will be right (except the last case) under the right circumstances. (The last case–isolated earth ground rods–is potentially very dangerous, and should never be done, particularly for medical devices.)

So the first order of business is to offer a common definition. A definition we like is widely used in the EMI design community, which says that a ***ground is simply a return path for current flow.*** These paths can be intended or unintended. The latter are often referred to as "sneak grounds," and can cause all kinds of EMI problems. Furthermore, a physical connection may not even exist as ground currents may hop onto an unintended ground path through parasitic capacitance or inductance.

When dealing with ground currents, it helps to remember some fundamental circuit concepts.

- A finite current flowing across a finite impedance results in a finite voltage drop (Ohm's Law). Thus, there is no such thing as a "zero potential" ground in the real world. Even microamps across microohms result in picovolts of potential drop.
- Currents must return to their source, and furthermore, these currents will take multiple paths with amplitudes inversely proportional to the impedances in these paths (Kirchoff's Law). Thus, unintended currents can and do flow in unintended paths.

Fortunately, unintended currents can be minimized and the unintended potential drops can be limited to acceptable levels, but only with understanding and proper design techniques.

DIFFERENT TYPES OF GROUNDS

Grounds are used for many reasons, including power, safety, lightning, EMI, and ESD. Although they share the common function of providing a return path, they differ widely when it comes to current amplitudes and frequencies. Recognizing these key differences are crucial to understanding grounding issues.

As stated in chapter 3, we can move between the time domain and the frequency domain using the EFFT, where $f_{eq} = 1/\pi t_r$. Thus, we can look at frequency dependence for time-domain transient events. Simply stated, the faster the edge rates, the higher the frequency content. For example, an ESD transient with a 1 ns rise time has an equivalent EMI frequency of about 300 MHz, while a lightning transient with a 1 μs rise time has an EMI frequency of only about 300 kHz. Why is this possible? Ben Franklin showed that lightning and ESD are both "electrical," yet their frequency characteristics differ by three orders of magnitude.

Table 6-1 shows some typical frequency and amplitude requirements of several different types of grounds. We see that the power and safety grounds must be able to handle high currents, but only at low frequencies. Grounds for EMI and ESD, on the other hand, need to handle lower currents but at very high frequencies. We also see that analog frequencies include the power frequency range but often operate at very low current levels, which is why analog circuits are often upset by 50/60 Hz power currents. Even a few microamps of power "leakage" current can completely smother intended low level analog signals.

When dealing with grounding issues, you must consider both the frequencies and the amplitude of ground currents. We will see shortly that frequency concerns also dictate ground topology, such as single-point grounds vs. multipoint grounds. The needs of these different types of grounds are different and must never be confused.

MICROSHOCK HAZARDS—A SPECIAL MEDICAL GROUNDING CONCERN

One of the reasons grounding gets so much attention in the medical world is the possibility of "microshocks" to patients in patient-connected equipment. This is usually due to "leakage current" where a small portion of the power return current takes an alternate path back to the source, rather than following the intended power neutral path. If that "leakage current" flows through a human, the results can be fatal.

Leakage currents occur in many electrical and electronic devices. This leakage results from the power line voltage (50/60 Hz)

Table 6-1. Frequency and Amplitude Requirements for Grounds

Function	Frequency	Amplitude	Duration
Safety	50/60 Hz	10 to 100 A typical	ms to hours
Power (neutral)	50/60 Hz	10 to 100 A typical	Continuous
Lightning	300 kHz	100 kA	50 ms
ESD	300 MHz	mA to A	50 ns
Digital	1–300 MHz	mA to A	Continuous
Analog	DC–100 kHz	μA to mA	Continuous
EMI	DC–Daylight	μA to mA typical	Continuous

being impressed across "line-to-ground" capacitance, which results in a small amount of current flow through the capacitive reactance. This capacitance can be due to "line-to-ground" capacitors in an EMI filter, or simply due to parasitic capacitance in the device itself. This is shown in Figure 6-1.

For nonmedical devices safety agencies such as UL or EU typically limit leakage currents to a few milliamps, because most humans can withstand these levels with no adverse affects. For patient-connected medical devices, however, the safety agencies limit the leakage currents to much lower levels, typically 20 μA. This is because medical devices usually have a much lower contact impedance with the human. In fact, devices like ECGs often use conductive paste to lower the contact impedance to facilitate measurements.

WHAT MAKES A GOOD GROUND?

A good ground must have a low impedance to minimize the voltage drop along the intended path and to provide a preferred path for current flow. *The key to success is maintaining that low impedance*

Figure 6-1. Microshock Hazards

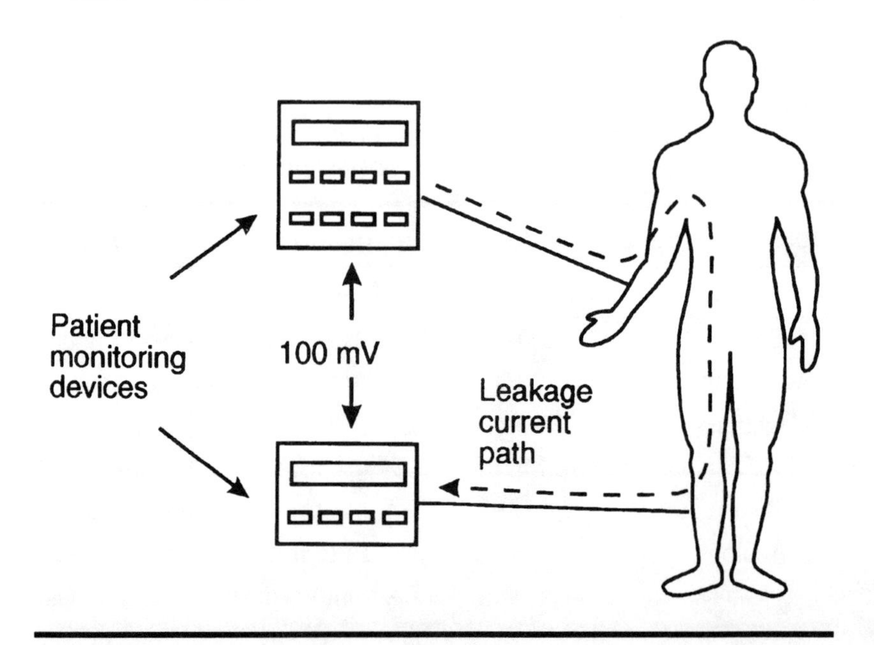

over the entire frequency range of interest. We cannot overemphasize this point. Most EMI grounding problems are due to using the wrong approach for a given frequency range.

The impedance (Z) of any ground conductor consists of both resistance (R) and inductance (L) ($Z = R + j\omega L$) For DC and low frequencies resistance is the major factor, so heavy gauge wires are often used for low frequency grounds. As the frequency increases, however, the inductance becomes more important. For most round wires this inductance is in the range of 8 nH/cm, so the inductive reactance becomes a factor at frequencies above 1–10 kHz range, depending on the wire size. This is shown in Table 6-2, which shows several parameters for different wire sizes: resistance (R) per meter, inductance (L) for 10 cm, and inductive impedance (Z) per meter at several frequencies.

> Example–The impedance of a 10 cm length of 22 gauge wire at 100 MHz is 81 Ω. This is far from the intended short circuit.

It is apparent that at power and audio frequencies (DC–10 kHz), resistance is the dominant factor in ground impedance. Thus, at low frequencies we look for ways to reduce the resistance, typically by using larger wires or multiple wires. At frequencies above the audio range (>10 kHz), inductance becomes the dominant factor in ground impedance. Thus, at even moderate frequencies we must we look for ways to reduce the inductance of the ground path. Two ways to accomplish this are to use wide straps or full ground planes. (A variation on the ground plane is to use a ground grid.)

Let us look closer at inductance in grounds. Inductance is merely a measure of the ability to store energy in an adjacent magnetic field. Round wires are quite efficient at this, and this property can be further boosted by winding the wire into coils, with the magnetic flux linking multiple turns. The inductance can also be increased by providing a permeable material to further concentrate the magnetic flux. Our objective, however, is not to increase inductance, but to reduce it as much as possible.

Ground Straps

One way to reduce inductance is to "flatten" a wire, or turn it into a strap. The increased surface area means that the inductance per unit length decreases. The amount of decrease depends on the aspect ratio (length to width), which determines how much the current can "spread out," as illustrated in Figure 6-2. The amount of decrease is shown in Figure 6-3, a graph taken from MIL-HDBK 419.

Table 6-2. Parameters for Wire Sizes

Gauge (Awg)	Diameter (mm)	Ω/meter	L/ten cm* (nH)	Inductive impedance (*Z*) of wire in Ω for 10 cm length		
				@ 10 kHz	@ 1 MHz	@ 100 MHz
10	2.5880	0.0033	101	0.006	0.63	63
12	2.0530	0.0052	105	0.007	0.66	66
14	1.6280	0.0083	110	0.007	0.69	69
16	1.2910	0.0132	115	0.007	0.72	72
18	1.0240	0.0209	119	0.007	0.75	75
20	0.8118	0.0333	124	0.008	0.78	78
22	0.6438	0.0530	129	0.008	0.81	81
24	0.5106	0.0842	133	0.008	0.84	84
26	0.4049	0.1339	138	0.009	0.87	87
28	0.3606	0.1688	140	0.009	0.88	88
30	0.3211	0.2129	143	0.009	0.90	90
32	0.2546	0.3386	147	0.009	0.92	92
34	0.2019	0.5385	152	0.010	0.95	95
36	0.1601	0.8564	156	0.010	0.98	98
38	0.1270	1.3609	161	0.010	1.01	101
40	0.0799	3.4410	170	0.011	1.07	107

*Calculated using $L = 12.9 \times \ell \times \ln(4\ell/d - 0.75)$

Figure 6-2. Current Spreads Out in Ground Strap

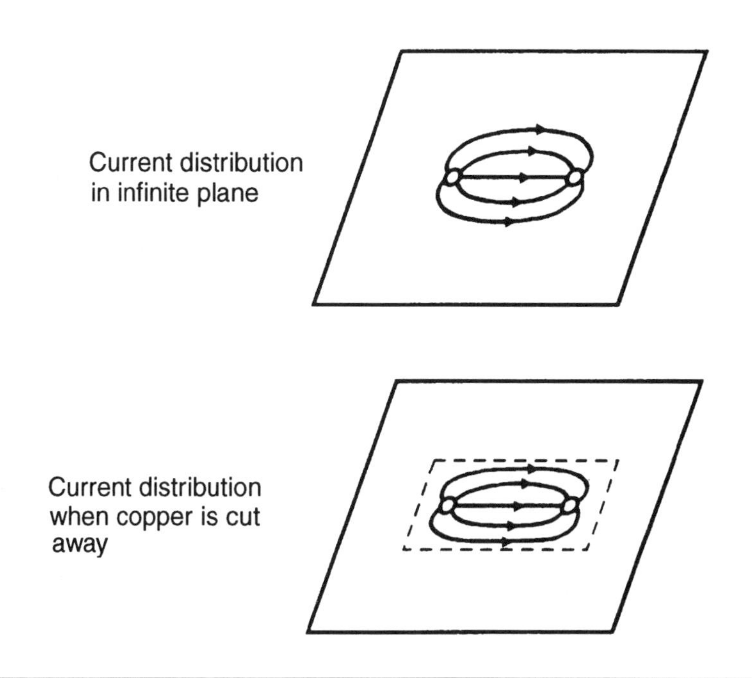

A widely accepted guideline in the EMI world (and the RF world) is to keep the length to width ratio less than 5:1, preferably less than 3:1. We prefer the 3:1 ratio ourselves, especially for high frequency grounding above 100 MHz. This includes ESD, with an equivalent EMI frequency of about 300 MHz. This means that a 3 cm ground connection must be at least 1 cm wide. A common problem we see is someone using 3 meters of 2 cm wide braid and assuming they have a "low inductance" ground. The fact is, with an aspect ratio of 150:1, you might as well use a regular round wire.

Ground Planes

The ultimate "strap" is the solid ground plane. In this case both the resistance and the inductance are greatly reduced, due to the parallel impedance paths. This is shown in Figure 6-4. Impedance of ground planes is usually given in "ohms-per-square," or the impedance across opposing sides of a square area. Note this dimension is unitless since the impedance decreases in parallel at the same rate as it increases in series.

Figure 6-3. Impedance of a Ground Strap Relative to a Wire (Source: MIL-HDBK 419)

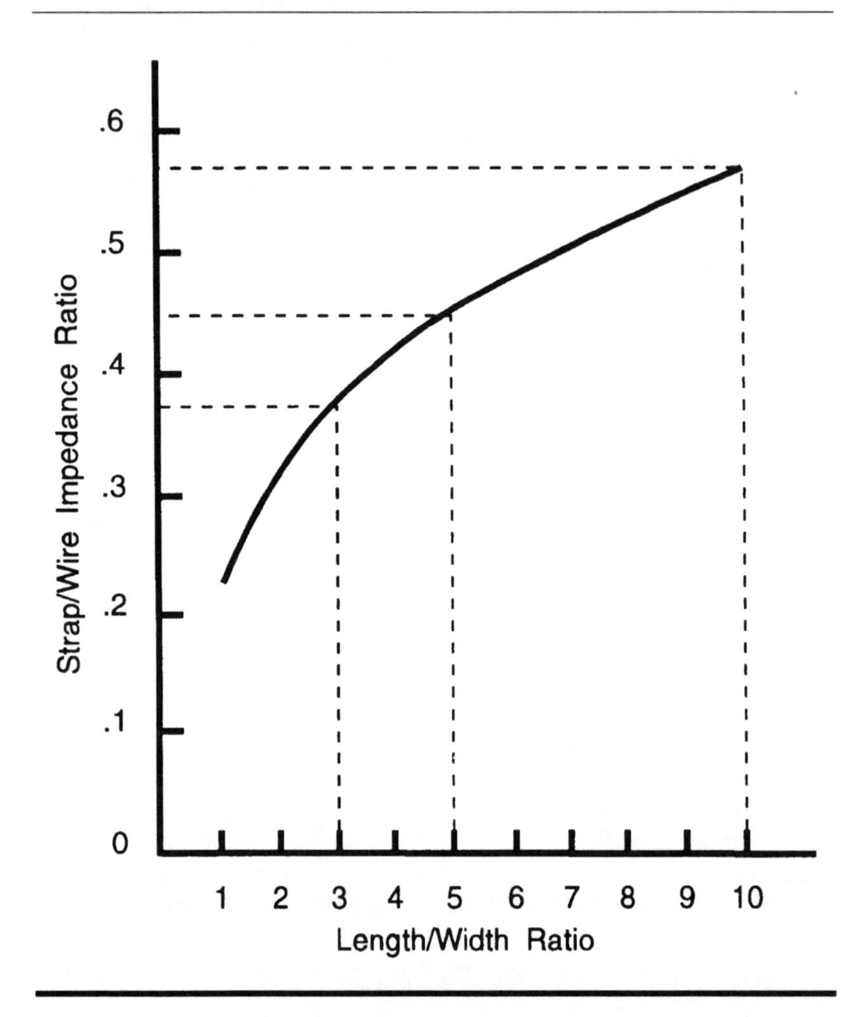

At low frequencies the thickness of the ground plane contributes to its resistance. That is, the thicker the material, the lower the resistance in "ohms-per-square." This is similar to using a larger wire at low frequency to lower the overall resistance.

As frequency increases, thickness becomes less of a factor, due to the skin effect. In fact, once the thickness of the ground plane exceeds about three skin depths, it is no longer a contributing factor. For example, aluminum foil represents about three skin depths at 100 MHz, so for frequencies 100 MHz and up, there is virtually no

Figure 6-4. Impedance of a Ground Plane Relative to a Wire

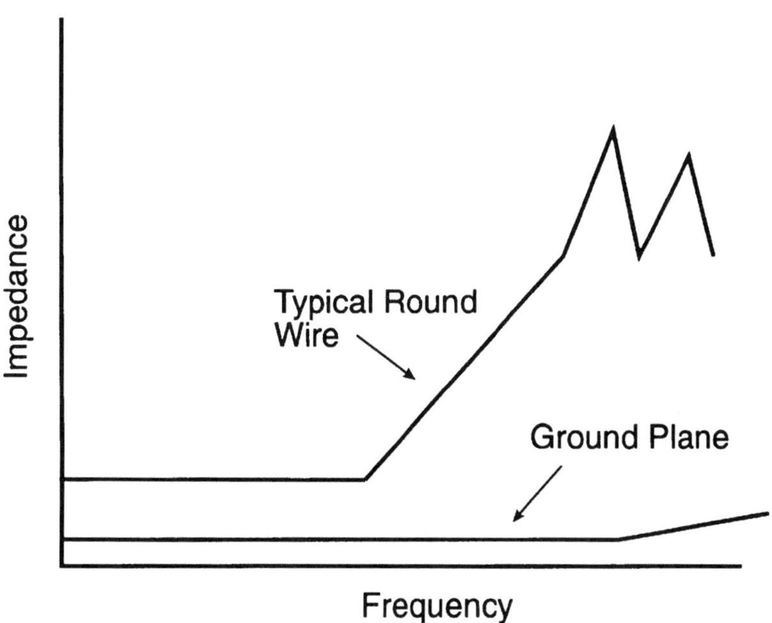

difference in ground impedance between aluminum foil and a 25 mm solid aluminum plate.

Table 6-3 shows some typical values of ground plane impedance for solid copper planes over a wide frequency range. As you can see, at power frequencies and audio frequencies the impedance decreases with thickness, but at higher frequencies, thickness becomes less important as the skin effect becomes a limiting factor. This is analogous to the inductance in a round wire (or even a strap), except that the skin effect increases as the square root of frequency, while inductance increases directly with frequency. This is shown in Figure 6-4, which also shows the effects of wire impedance with frequency.

There are two points to be made here:

1. At frequencies above about 10 kHz, planes provide much lower ground impedances than wires. At 100 MHz this can be several orders of magnitude.

Table 6.3. Values for Ground Plane Impedance

	Thickness		
Frequency	**0.1 mm**	**1 mm**	**10 mm**
60 Hz	172 μΩ	17.2 μΩ	1.83 μΩ
1 kHz	172 μΩ	17.5 μΩ	11.6 μΩ
10 kHz	172 μΩ	33.5 μΩ	36.9 μΩ
100 kHz	175 μΩ	116 μΩ	116 μΩ
1 MHz	335 μΩ	369 μΩ	369 μΩ
10 MHz	1.16 μΩ	1.16 μΩ	1.16 μΩ
100 MHz	3.69 μΩ	3.69 μΩ	3.69 μΩ
1 GHz	11.6 μΩ	11.6 μΩ	11.6 μΩ

2. As frequencies increase, even solid planes become frequency limited due to skin effects. The increase is slower than with a wire, however, as plane impedance increases as the square root of frequency.

Ground Grids

Although a solid ground plane provides the ultimate in high frequency grounding, this is not always possible or practical. Nevertheless, the many benefits of a ground plane can often be achieved by a ground grid. The secrets of success here are to limit the inductance by limiting the space between the grids, and to make both the ground grid and the interconnecting elements as "fat" as possible.

A generally accepted criteria for grid spacing is 1/20 of a wavelength, based on the highest frequency that the grid is expected to handle. This is not as severe a constraint as it first sounds, for even the typical 2-foot spacing (60 cm) used on computer room floors to form a "signal reference grid" is good to about 25 MHz. This is usually more than adequate for computer systems, where the objective is to control power transients and noise typically in the 1 MHz range and below. The 2-foot spacing (60 cm) is not adequate, however, for ESD transients in the 300 MHz range.

A 1/4-inch (65 mm) mesh is often used for ground planes at EMI test facilities. This spacing is good to about 2 GHz, which is more than adequate since most EMI testing only goes to 1 GHz. The gridded ground concept also works well on PCBs. A 1/10-inch (2.5 mm) spacing, typical on circuit boards, is good to over 6 GHz by this criteria.

GROUND TOPOLOGIES

Now that we have looked at ground impedance vs. frequency, we need to look at ground topologies vs. frequency. This is very important, since we will soon see that for low frequencies (typically under 10 kHz), single-point grounds are the preferred approach, while for high frequencies (above 10 kHz), multipoint grounds with ground planes are the preferred approach.

Since most medical devices include sensitive analog circuits operating at low frequencies, and noisy digital circuits operating at high frequencies, both approaches may be necessary in the same device. The secret is knowing which type of ground to use in each situation. In some cases we even resort to "hybrid" grounds, using capacitors and inductors to alter the ground topology as the frequency changes.

Single-Point Grounds

At low frequencies we can usually steer ground currents via wires. The inductance is low, so the limiting factor is the wire resistance itself. Furthermore, capacitive coupling from the wires to adjacent wires or surfaces is small, so virtually all the current flows in the ground wires themselves.

A typical grounding scheme is shown in Figure 6-5. Note what happens if the system is "grounded" on both ends. Any common noise current in the common ground path is now coupled into the circuit via the "common ground impedance." This is a very common problem in low frequency applications–known as the "ground loop," which will be discussed in more detail a bit later in this chapter.

One way of preventing any common ground currents from getting into our circuit is to establish a "single-point" ground, as shown in Figure 6-6. In this case there is no "common impedance" across which we can generate a common voltage due to another system's ground currents. Thus, single-point grounding is a very practical way to limit "ground noise" problems in circuits and systems in which low frequency ground currents from other circuits are a threat.

Figure 6-5. Typical Grounding Scheme

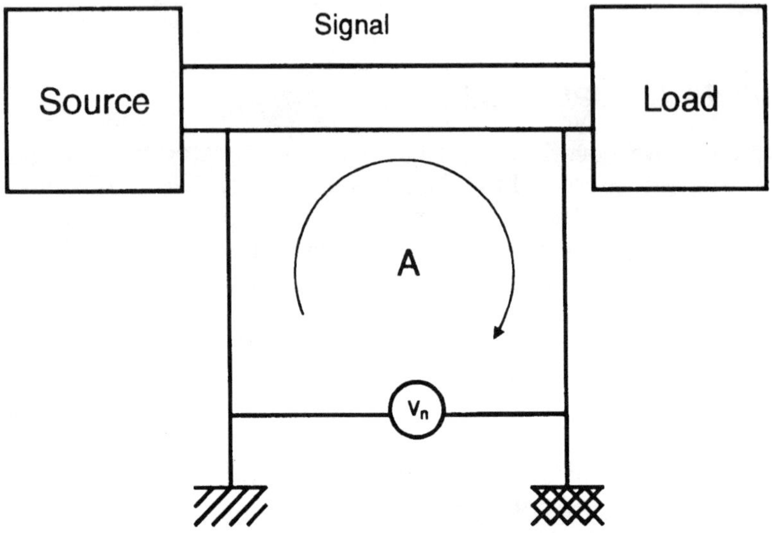

Multipoint Grounds and Ground Planes

Unfortunately, as the frequency increases, the inductive reactance of the wires increases, the parasitic capacitive reactance decreases, and we reach a point where we can no longer maintain a "single-point ground," no matter how hard we try.

The high frequency model of a single-point ground is shown in Figure 6-7. As the frequency increases, the current is more likely to flow through the capacitance rather than the inductance. For example, let us assume a 10 cm wire connection at 8 nH/cm, or 80 nH, and a parasitic capacitance of 100 pF. At 300 MHz (ESD frequencies), the reactance of the parasitic capacitance is about 5 ohms, while the inductive path is about 50 ohms. Clearly, most of the ESD transient current follows capacitive path, not the inductive "single-point ground" path. The same is true for any VHF radio frequency energy, for both emissions and immunity.

Thus, for high frequencies it appears that single-point grounds do not work very well. Since the current is going to couple to nearby surfaces, why not just "bolt it down" anyway, and then work on decreasing the impedance of the ground surface. The latter, of course, calls for planes or grids.

Figure 6-6. Single-Point and Daisy-Chained Grounds

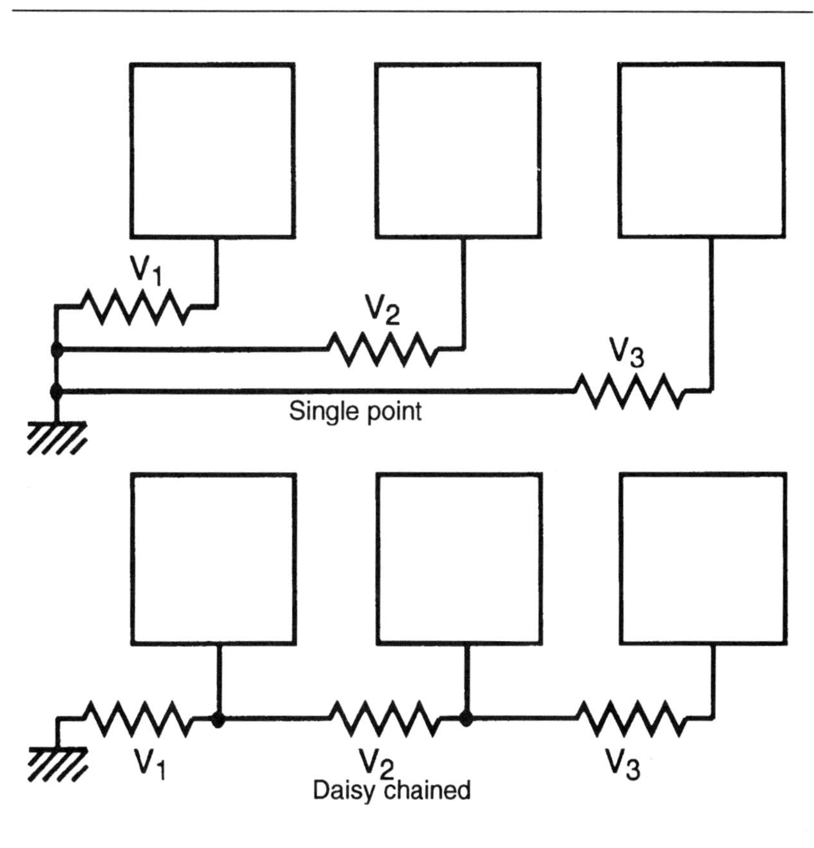

It is generally not adequate to use a single-point connection to a grid or plane. Rather, multiple connections are favored for high frequencies, due to transmission line effects. Remember, a "short circuit" becomes an "open circuit" at a distance of 1/4 of a wavelength. Standing waves are common on longer cables, as shown in Figure 6-8.

> For example, at 300 MHz a wavelength is 1 meter, so 1/4 of a wavelength is 25 cm. At 1000 MHz, that distance decreases to only 7.5 cm. The idea is to have multiple connections to limit these transmission line effects.

The recommended spacing for multipoint grounding is 1/20 of a wavelength or less, the same as for spacing of ground grids

Figure 6-7. High Frequency Model of Single-Point Ground

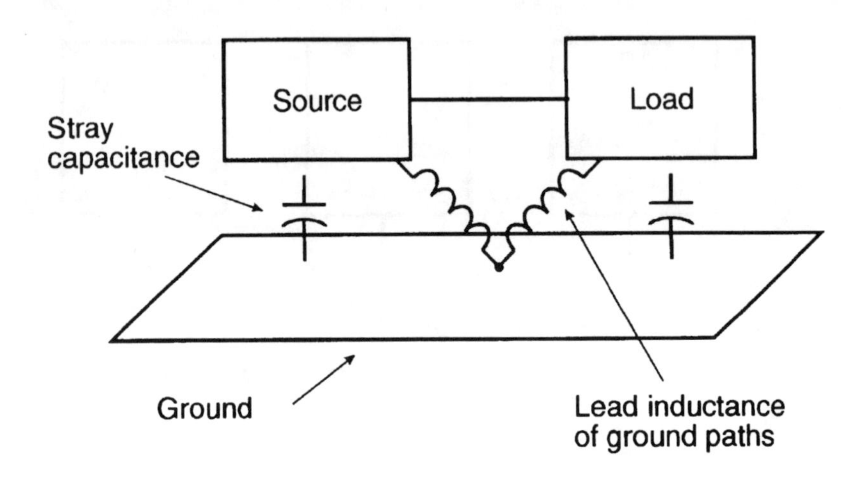

Figure 6-8. Standing Waves on Cable

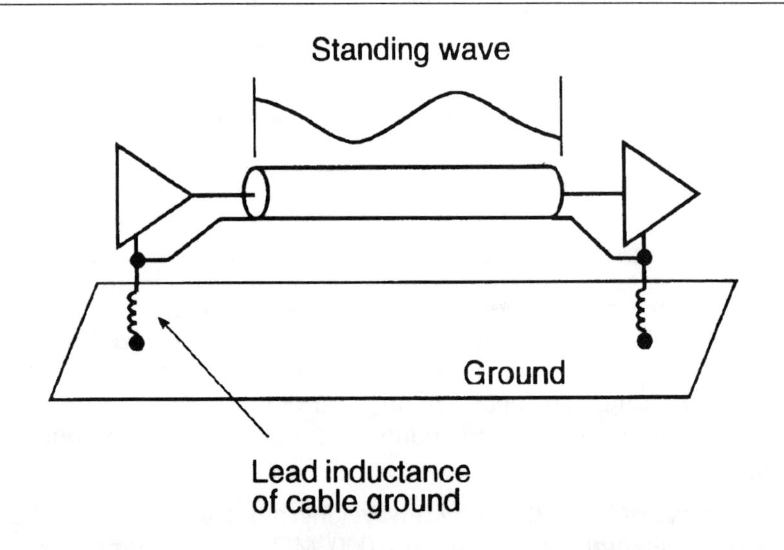

mentioned earlier in this chapter. This means about every 5 cm for frequencies up to 300 MHz.

Where you ground is as important as how you ground at high frequencies. This is particularly true for cables that terminate on circuit boards. Figure 6-9 illustrates the problem of using a single-point ground on a PCB with a cable shield terminated on the opposite end of the board. At high frequencies the transmission line effects of the board actually form a coupling network that drives noise currents onto the cable shield, causing it to radiate. This unwanted network is very similar to the "gamma match" used on UHF and VHF antennas.

The secret to success with cables is to ground both the cable and the board at the same point. In effect, this "shorts out" the "matching network," and prevents high frequency noise currents from being coupled onto the cable. We will look at cable shield grounding in more detail in chapter 10.

Hybrid Grounds

Sometimes our grounding system must serve two masters–high frequency and low frequency. How do we handle this situation? We resort to hybrid grounds, using capacitors and inductors to steer the appropriate frequencies in the desired directions.

Figure 6-10 shows an example at the circuit board level, which is popular with personal computer manufacturers. In this case all the ground pads but one are capacitively coupled, with one having a DC connection as well. Thus, at high frequency the system has a multipoint ground, while at low frequencies a single-point ground is maintained.

Figure 6-11 shows an example of grounding a cable shield on an analog circuit. The shield is connected to the cabinet through a

Figure 6-9. CM Radiation from a Single-Point Grounded PCB

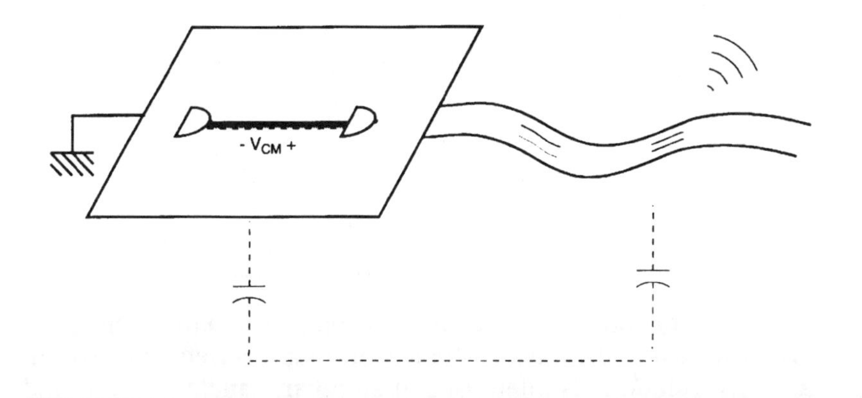

Figure 6-10. Hybrid Ground on PCB

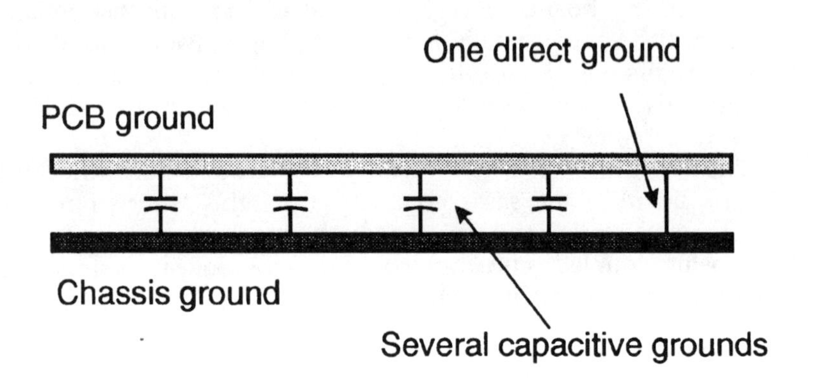

Figure 6-11. Analog Cable Shield Ground

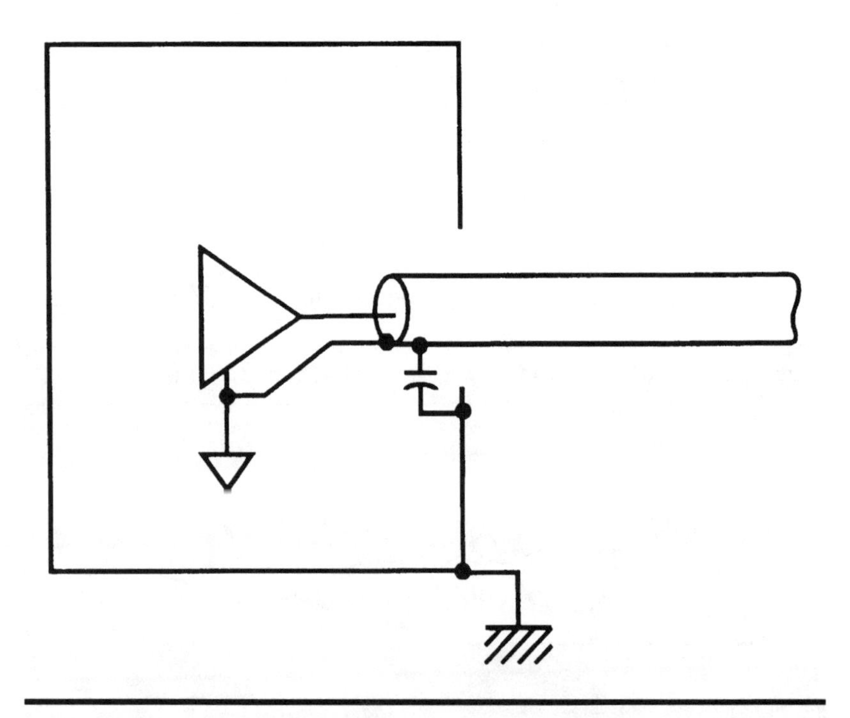

capacitor to provide a high frequency termination to the cabinet, yet it is connected to the analog circuit return for low frequency shielding. This technique is often used in audio and analog circuits and LANs. In fact, in the latter case, special BNC connectors with built-in capacitors are available from several vendors to provide a "hybrid" ground termination for LAN coaxial cables.

A choke in the "safety ground" lead is another type of hybrid ground. In this case 60 Hz is passed, but higher frequencies (RF and transients) are blocked. This optional choke is popular in commercial EMI power line filters. A few words of caution on this approach. First, this choke may cause other problems with ESD. Second, too much inductance can limit 50/60 Hz fault currents. For this reason we only recommend doing this with approved EMI filters—we *do not* advocate this approach with separate, external inductors.

Ground Loops

Ground loops are a serious problem for sensitive analog circuits facing low frequency threats. At high frequencies ground loops generally do not pose serious threats if proper high frequency precautions are taken in designing the ground system.

A ground loop exists whenever multiple ground paths exist. The problem is that unwanted currents may take an unexpected path, resulting in unwanted noise voltages interfering with intended voltages. The problem is particularly acute in low level analog systems, where even microvolts of noise can jam the intended signals. A classical example is where 60 Hz return currents pass through an audio system, causing a loud hum.

Figure 6-5 shows a typical ground loop problem. Note that it takes three conditions for a ground loop problem to occur. First, there must be a common shared impedance. Second, there must be an unwanted source of current. Third, there must be a circuit that is vulnerable to the levels of voltage caused by the unwanted currents. As with any EMI problem, you need a source, a path, and a receptor.

Since we usually cannot do anything about the sources or receptors, we attack ground loop problems at the path level. As we have already seen, single-point grounds work very well at low frequencies, where we can readily steer the current. At high frequencies parasitic capacitance and inductance creates multiple paths anyway, so we accept the "ground loops" and work to lower the common impedance (and thus the resulting voltage) with planes or grids.

If we cannot change the grounding points, we can still attack the ground loop by "breaking" it in other places. Ground loops cause common mode (cm) currents to flow, so providing cm isolation anywhere in the loop can control the problem. Figure 6-12 gives several examples of breaking ground loops. In addition to ground isolation,

Figure 6-12. Eliminating a Ground Loop

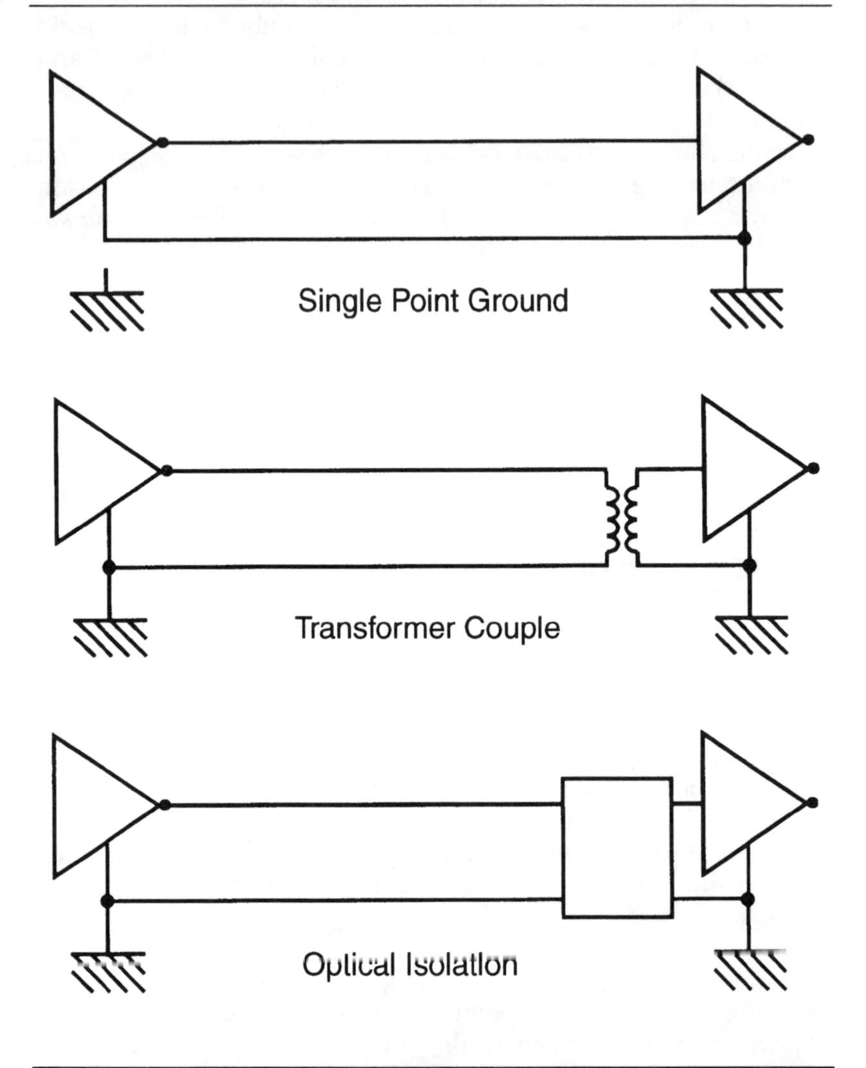

we can insert either transformers or optical isolation in the signal path. All will block cm currents, while allowing differential mode (DM) currents to pass. Optical isolators and fiber optics are also very effective at breaking ground loops. Fiber optics also have other advantages (no antenna affects on cables), so they are seeing increased use in harsh EMI environments. Transformers will also block differential DC, which may or may not be desirable.

Isolated Grounds and Earth Grounds

Often someone will suggest an "isolated" ground to solve a vexing EMI problem. At this point we become very concerned, since an improper "isolated" power safety ground can result in a very unsafe system.

Isolated power grounds are another version of the single-point ground. The idea is to prevent unwanted currents the opportunity to hitch a ride on your power ground. Power circuits with isolated grounds are often used in medical environments, and are identified by special orange power outlets. In this case, the green wire safety ground is physically isolated from the conduit and other building steel, *but a dedicated grounding conductor is carried back to the power source.* This allows the circuit breakers to clear if a fault (short from power to case) occurs.

The improper, and unsafe, isolated ground is implemented by using a separate "ground rod" that is not connected to the power source, but relies on a local earth connection. Sure, it may solve the EMI problem, but if a power fault ever occurs, the impedance back to the power source may be too high to open the circuit breaker. This is a very dangerous possibility and must be avoided. It also violates the National Electrical Code, so *don't ever do this . . . ever!* (Unfortunately, we still see this from time to time–it is not only dangerous, it is irresponsible engineering.)

Figure 6-13 shows some examples of isolated grounds, both good and bad. One other comment on earth grounds. You generally only need these for power safety and lightning. How many airplanes use earth grounds at 30,000 feet? How many laptop computers require an earth ground? Remember, ground is a path for current flow–not a "dumping ground" for unwanted current. So, do not get hung up on earth grounds, particularly for electronic systems.

BONDING CONSIDERATIONS

By now it is apparent that maintaining low impedances in ground systems is crucial. A very important part of this is the joints or connections between parts of the ground system, be they wires, planes, grids, or straps. Unless the joints are welded, corrosion can form and cause high impedance points in the ground system. This is particularly important with medical devices, since they are often subjected to chemicals or body fluids.

Two types of corrosion can occur–galvanic and electrolytic. In the galvanic case two dissimilar metals act as a miniature battery,

Figure 6-13. Isolated Grounds

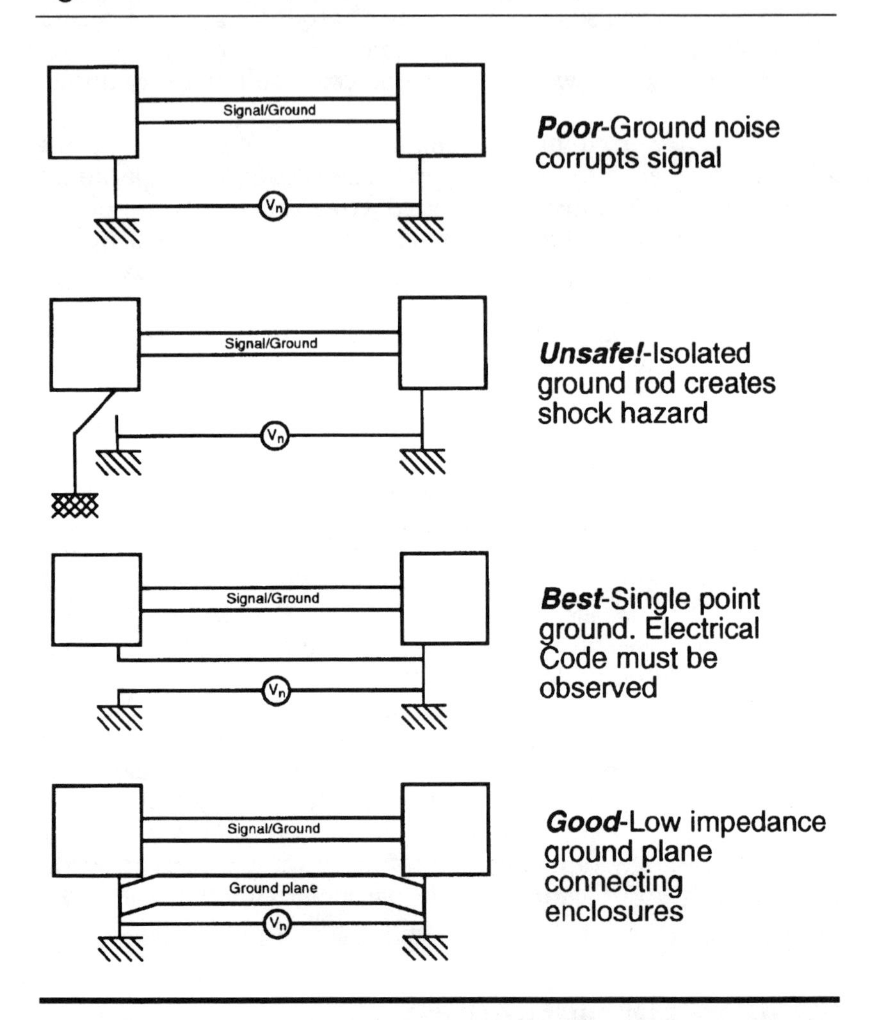

resulting in a small current flow that eventually corrodes the metal at the connection. In the electrolytic case external currents cause the corrosion. Since both cases require moisture, sealing the joint prevents corrosion.

If you cannot seal against moisture, then metals close to each other in the electrochemical series will help minimize corrosion. This series is shown in Appendix B. Metals that are far apart in the series can be plated so that the exposed mating surfaces are the same material, or at least closer in the series than the base metals.

GROUNDING DESIGN GUIDELINES

By now it should be apparent that there is no single "magic bullet" for grounding. Rather, different techniques and methods are necessary for different types of circuits and operating conditions. Here are some specific design guidelines for different types of electronic circuits. Keep in mind that these guidelines may still need to be modified for a particular situation, but these provide a good starting point for your designs.

Digital Circuits

The key grounding concerns for digital circuits are containing emissions (radiated and conducted noise in the RF range), maintaining signal integrity (crosstalk, ground bounce, etc.), and withstanding external noise like ESD and RFI from radio transmitters.

Since most digital circuits operate at high frequencies, multipoint grounds and ground planes are preferred. The connections between the circuit and its grounds should be short, fat, and direct. Remember wires and traces are highly inductive (8 nH/cm typical), so straps or other "fat" connections such as gaskets are preferred.

On a multilayer circuit board the power and power return planes both act as local high frequency ground planes. The circuit board return plane should also be connected to a metallic enclosure through multiple low inductance connections. If the enclosure is plastic, you may want to consider adding a metal plate under the circuit boards. This approach is popular in commercial electronics (printers, fax machines, etc.) where full shielding is not needed.

On two-layer circuit boards, you have no local high frequency ground planes. For this reason we strongly prefer multilayer boards for digital circuits, as they are generally 10–100 times quieter than two-layer boards. If you insist on using two-layer boards, multipoint connections to the cabinet can be beneficial.

Analog Circuits

The key grounding concern for analog circuits is jamming of the analog circuits. This is particularly critical for low level circuits. If you are looking for a 1 μV signal, and you have 1 mV of ground "noise," you have a big problem. And if your noise is in the frequency range of interest, you cannot rely on filtering to remove it, which is typical of most low frequency analog sensors operating in the DC–10 kHz range.

Since most analog circuits operate at low frequencies, single-point grounds are preferred. The most typical ground current threats to analog circuits are stray 50/60 Hz power return currents, stray switching mode power supply return currents, and digital circuit return currents. As a result, make sure you have only single-point connections to any of these circuits. In systems using A/D and D/A converters, you must pay special attention to the analog/digital connection. This is discussed in more detail in chapter 10.

Higher level analog circuits are not as vulnerable as low level circuits. Our general criteria is to limit ground voltages to less than the smallest operational increment of the circuit. For example, if we are using a 10 bit A/D converter, the least significant bit represents about 5 mV in a 5 V system. Thus, a millivolt of noise probably is not a problem. You can relax this even more if you do not care about real fine resolution. On the other hand, if you are looking for a 10–100 μV signal, 1 mV of noise is way too much.

I/O Circuits

The key grounding concerns for I/O circuits are both emissions and immunity. Since the I/O connects to the outside world, this is a critical point for blocking EMI into or out of the system. Proper grounding plays a very important role here, and, of course, leakage current limitations may significantly alter these guidelines. Figure 5-15 shows the preferred grounding for signal line filters.

For *immunity* against outside threats like ESD and RFI, we recommend grounding bypass capacitors or transient protectors directly to the cabinet. If a cabinet is not used, or it cannot be used to terminate a capacitor filter, then a separate metal plate under the circuit board should be considered. The objective is to keep RF or transient currents from entering either the circuit or the circuit ground.

If you do not have an enclosure or other chassis, then you will need to connect to the circuit ground, and this is a decidedly less satisfactory choice. In this case we recommend a series element (resistor or ferrite) to limit the current as much as possible. This will help minimize the "ground bounce" due to the external current. You never use a bypass capacitor to shunt external interference currents directly to the signal ground, analog or digital–always insert some series impedance first.

For *emissions* due to currents originating at the chip, we recommend grounding bypass capacitors directly to the appropriate signal ground. In this case the objective is to return these currents to their respective sources. A series element helps here, as it limits the amount of interference current exiting from the chip. For emissions due to common mode currents generated onboard, this approach will not work–you need to shunt the currents to the chassis, the same as with external interference.

If you are dealing with both emissions and immunity, you may need two shunt elements, with one grounded to the signal ground for emissions and one grounded to the chassis for immunity. The two shunt elements should be separated by a series impedance, such as a small resistor or ferrite. See Figure 5-15 for an example of I/O grounding.

Power Safety Grounding

The key concern for power grounding is human safety. If your system can be powered by 120 V alternating current (AC), then any exposed metal must be bonded to the "green wire" safety ground on the power plug. This also applies to battery-operated devices if the charger is built-in the unit–that is, if it can be charged from 120 V AC. If, on the other hand, your equipment is powered by modular plug-in power supplies that only provide 12 V DC to the unit, the safety grounding applies only to the power supply itself. You still need to meet the appropriate leakage current requirements, however.

A final piece of advice here. If there is ever a conflict in EMI and power safety grounding, the safety grounding shall always prevail. No exceptions to this rule is ever justified by any type of equipment, medical or otherwise. There are far too many case histories of electrocution to take this issue lightly. If you have been smugly doing this on the sly, shame on you: your practice may kill someone.

ESD and Static Grounding

Although we already discussed I/O grounding, some extra advice is in order for ESD in isolated circuits. Due to the leakage current limitations on patient-connected devices, the input circuits are often completely DC isolated from the rest of the device. Under ESD test conditions voltage levels can build up and cause a secondary arc that can upset or damage the device. To prevent this type of problem, you may want to add a very high impedance resistor (22 meg or more) between the isolated circuit to the rest of the circuitry to "bleed" any charge buildup. Keep in mind this will add to your leakage current, so if you are close to the limits, this may not be possible without other circuit changes. As an alternate, you could add an arc device between the circuits, to allow you to at least control the secondary breakdown path. A third option is to determine where the breakdown occurs, and add some damping with a ferrite or resistor. Although this is the least desirable approach, we have done this on existing products as a remedy that works.

SUMMARY

Grounding is a most misunderstood concept, the problem arising from the multiple purposes to which a ground may be used. Most of the old lore about grounds arises from power/safety purposes, followed by analog grounding techniques. Accordingly, the following checklist is offered to steer you in the right direction.

- A ground is a return path for current. Earth ground is not needed unless it is needed to return current to its source, as with lightning. Otherwise, a low impedance return path to the source is appropriate.
- A ground needs to be low impedance at the frequency of interest. With inductances of wire being 8 nH/cm, wire becomes too high an impedance for radio frequencies. Use a strap of 5:1 length to width ratio or lower.
- Single-point grounds are appropriate when the wavelength of the highest frequency of interest is longer than 1/20 the length of the interconnect. Otherwise multipoint grounds are appropriate.
- Hybrid grounds may be used as a compromise between multipoint grounds at high frequencies and single-point grounds at low frequencies.
- Long-term bonding integrity of mating surfaces needs to be ensured.

7

POWER AND POWER FILTERING

Power supplies are both a source and a conduit for interference. Power supplies generate noise from rectifier switching and snap-off and from the switcher or regulator. Power supply filters are usually designed for frequencies below 30 MHz, often ignoring the high frequency power supply noise, which is often significant up to 50 MHz, and completely fail to filter higher frequency system noise.

Power supply design for medical devices is a specialized field, and there some companies specializing in this discipline. This chapter is not intended to be a design guide for power supplies, but is intended to provide insights as to the EMI impact on power supplies.

POWER SUPPLIES

Power supply noise, like almost all noise, starts out as differential mode (DM), but coupling paths are almost all common mode (CM). This is extremely important when designing power supplies for patient-connected medical equipment, since leakage current limitations make CM filtering extremely difficult. Accordingly, there are two goals in power supply design: (1) block interference immediately at the source before it can become CM and (2) position your components so as to minimize coupling paths.

Power supplies generally fall into one of three categories: linear, ferroresonant, and switching. Whichever the technology, the tasks revolve around rectification and conversion to the desired supply voltage. Conversion usually involves regulation as well.

Linear Supplies

Linear supplies are the simplest. Start with a transformer to provide a voltage output of sufficient amplitude to provide adequate voltage for the worst-case conditions. The output is fed into a rectifier (usually full wave), passed through a ripple filter, then to the linear regulator. The regulator is provided the task of providing the correct voltage under all load conditions, and must either absorb or divert the difference between the unregulated supply and the regulated load voltage. Thus, if we have unregulated DC of 15 V source and a load of 12 V and 1 A, the regulator must either absorb or shunt 3 watts.

Linear supplies are very quiet from an EMI viewpoint, but do have their downside. Efficiencies are typically in the 30 percent range, meaning not only that the electronics are wasting 70 percent of the power, but also that the supply itself must dissipate that amount. Thus, linear supplies tend to be quite large for their output. Additionally, linear supplies are not typically very forgiving of power sags and surges.

Ferroresonant supplies

Ferroresonant supplies insert a resonant circuit in the line transformer, then feed a rectifier circuit and ripple filter. These devices are quite good for both line and load regulation, but are very frequency sensitive. This is a problem where frequency is not well controlled, such as a portable device or in underdeveloped countries. Efficiencies are about 75 percent. The output looks rather like a square wave, so ripple filtering tends to be rather minimal.

These devices operate on high magnetic flux densities (10 kG), which will adversely affect some nearby electronic devices, notably ballistic electron devices such as CRTs and electron microscopes. Ferroresonant devices are not widely used in electronics, but will be found in microwave ovens and in battery chargers, including uninterruptible power supplies.

Switching Supplies

Switching mode power supplies have become, in recent years, the work horse of modern electronics. Switching supplies consist of two basic elements, the AC to DC conversion, followed by the DC to DC conversion. In its most elemental form the AC to DC conversion starts by rectifying current directly from the power line, usually without an intervening transformer. A large capacitor serves as a coarse ripple filter. The rectified DC is then converted to the desired DC output voltage with the aid of a switching oscillator. The output may go directly to a load filter, or it may be transformed to a more suitable voltage before rectifying.

Load voltage is regulated by controlling the duty cycle of the oscillator. Thus, the regulator needs only to absorb the energy consumed during switching along with a small amount dropped through during on time. This means that the regulator wastes very little energy. Typical efficiencies are 80–90 percent.

There are several additional bonuses. First, the supply can be designed to tolerate wide ranges of line voltage and frequency variations. In fact, it is increasingly common to find supplies to operate on U.S. 120 V AC 60 Hz and European 240 V AC 50 Hz without setting internal switches. Typical domestic switching supplies will cheerfully operate on sags to 90 V AC.

Second, the switch operates at significantly above line frequency, typically 20–50 kHz, and the trend is to even higher frequencies. This means that the transformer size and ripple filter capacitor sizes can be reduced substantially.

As a result, switching supplies are far smaller and more efficient than linear supplies. But there is a downside, too. Switching supplies generate significantly more EMI. In fact, this is the principle reason that linear supplies are used in sensitive analog applications.

SWITCHING TOPOLOGIES

We are often asked which topology is the best from an EMI standpoint. Our observation is that there is not a large difference between them, so your decision should be for reasons other than EMI. Nevertheless, there are differences, and you may want to take advantage of them.

There are two major considerations when looking at a topology. The first is where the smoothing inductor is placed relative to the power source or load. The second is how much high frequency current is passed through the switch.

If the inductor is in series with the load but not the power source (buck converter), then the power source current will switch from full on to full off at the switcher rate and the load current will be smoothed by the inductor. If the inductor is in series with the load but not the source (boost converter), then the load current will switch from full on to full off at the switcher rate and the source current will be smoothed by the inductor. The buck-boost converter has no series inductor at all (it is in shunt), and so both input and output currents experience full switching. Finally, the Cuk converter (also called the boost-buck converter) has inductors in series with both the input and output, so both input and output currents are smoothed. These topologies are shown in Figure 7-1. Obviously, any

Figure 7-1. Switching Supply Topologies

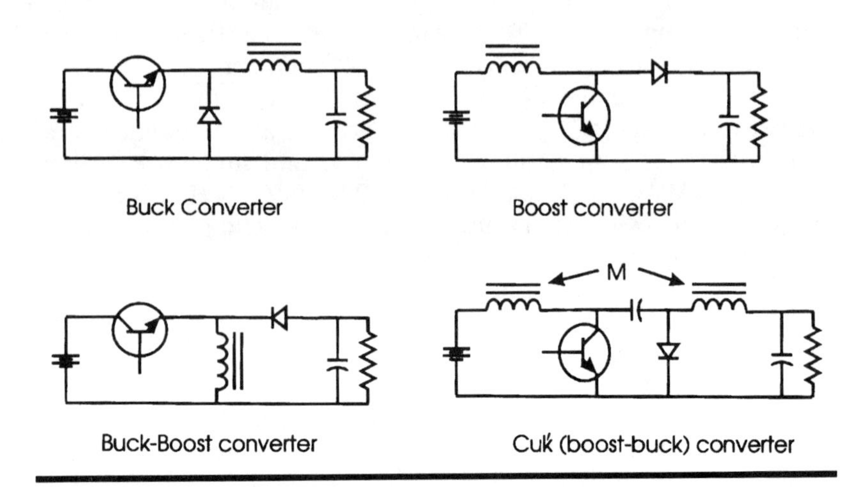

smoothing not provided by the converter would need to be accommodated either by filtering or by tolerating the ripple.

But note that these aspects are primarily low frequency issues. Even if the low frequency components are handled by topology or by supplemental filtering, there is still a need for high frequency filtering, either internal or external to the converter. As has been mentioned, intercepting interference as close to the noise source as possible is very desirable.

INTERFERENCE IN POWER SUPPLIES

Interference originates from the rectifier set and in the switcher (Figure 7-2). The rectifier generates low frequency rectifier interference as a result of drawing current from only the center of the cycle, thus generating low frequency harmonics. Additionally, the rectifier exhibits snap-off action, as a result of time needed to sweep out stored charge. Finally, some power supplies use thyristers instead of rectifiers for regulation, resulting in an abrupt voltage and current transition.

Obviously, these interference sources can conduct out the line, either to the power line or to the load. And the lower frequency components do just that. But the higher frequency components find alternate paths, as shown in Figure 7-2: capacitive coupling to structural members, capacitive coupling chopper to heat sink, capacitive

Figure 7-2. Interference Sources in Power Supply

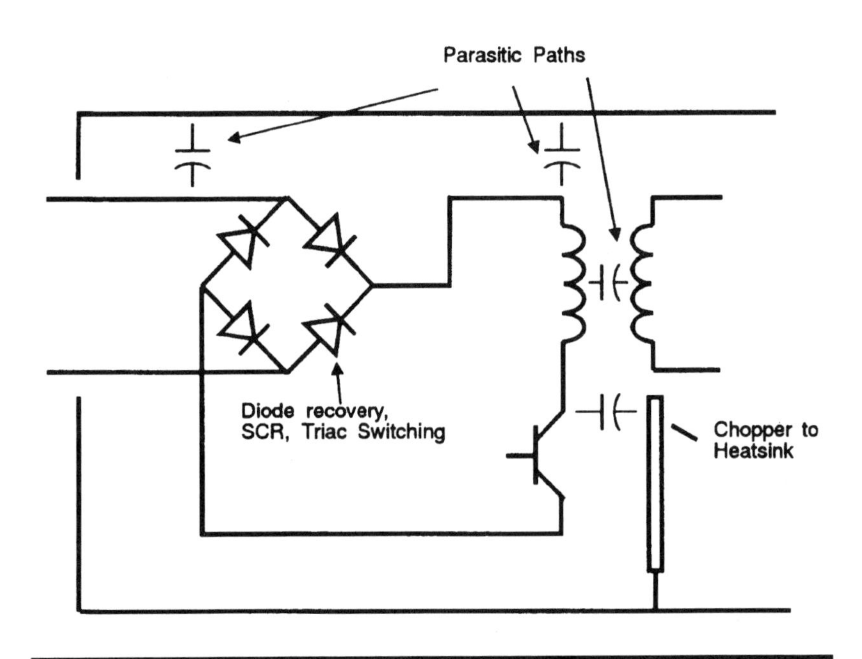

coupling to transformer, inductive coupling (loop) to any conductors. These parasitic paths may, in turn, couple back to the power line or load.

The interference sources are DM, as usual, so the low frequency conducted interference is differential. But the coupling paths are, by and large, CM, so higher frequency interference components tend to be CM. This is a key issue in medical electronic devices.

Common mode interference is most efficiently filtered by shunting currents to ground via a capacitor. If shunt capacitance is restricted, as it is in patient-connected devices, then the remaining option is to provide high series impedance in the form of an inductor. High frequency/high impedance filtering is extremely difficult to achieve, as parasitic capacitance will provide low impedance alternatives.

Accordingly, it is imperative to intercept the interference currents to the greatest extent possible before they become CM, and to lay out your power supply so as to minimize CM coupling paths.

Figure 7-3 shows how the coupling may be minimized. Capacitive coupling to enclosure is minimized by careful component

Figure 7-3. Minimizing Coupling Paths in Power Supply

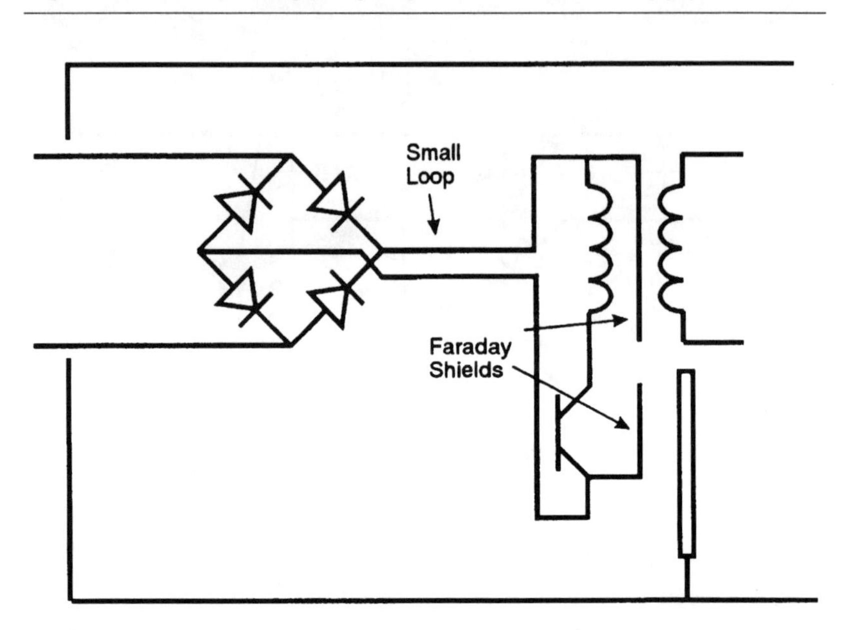

placement. Capacitive coupling to heat sink is avoided by installing a Faraday shield (such as Bergquist Sil-Pad®) or by isolating the heat sink. Capacitive coupling to transformer is avoided by a Faraday shield in the transformer. Inductive coupling is minimized by eliminating the loop areas by close spacing and preferably by twisting wires.

POWER FACTOR CORRECTION

Rectifier-derived supplies gulp current at the peak of the AC cycle (Figure 7-4), leaving the rest of the cycle unused, and generating power line harmonics in the process. The resulting harmonics result in increased heating on the power lines and in power transformers and, if the draw is sufficient, to adversely affect the voltage waveform, will affect the operation of three phase motors.

Accordingly, the EU will soon mandate power factor correction in electronics power supplies to minimize this effect. Power companies in the U.S. are looking closely at this same issue.

The goal is to distribute the power draw over a larger part of the power cycle, by spreading out the current draw over the cycle. In

Figure 7-4. Power Line Harmonic Generation

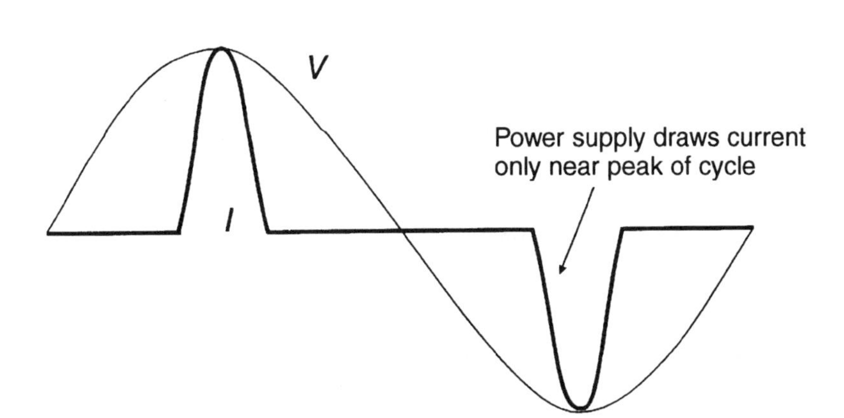

the simplest solution the output of the rectifier is very lightly filtered, leaving a large ripple that, in turn, is handled by the regulator downstream. The voltage output, of course, will be limited by the available voltage at the lowest voltage of the ripple, so steps must be taken to step up the voltage to adequate amounts.

Generally, a two-step conversion is used, with one used for source regulation followed by one used for load regulation. In fact, such practice is beneficial even when power factor correction is not needed, as it improves overall regulation. Whichever method is used, there will be copious amounts of high frequency components to be filtered out.

CONTROLLING INTERFERENCE GENERATION

It is important to identify the source of the interference and attack it as close to the source as possible, either at the rectifier set or at the switch. In the case of the rectifier snap-off noise, the problem can be minimized by using diodes with low stored charge.

Converter Interference

In the case of the converter, the interference source is the switch itself, which generates high time-rate-of-change of both voltage and current. Switcher edge rates are usually designed to be as fast as possible, as the faster the switch, the lower the power draw in the switching device itself. Fast switchers are the major source of interference. If you slow the switch, you reduce interference. Can we reduce interference without appreciably affecting function of the converter itself? In many cases we can.

The answer lies in the frequencies involved. Suppose we have a switcher running at 50 kHz. For good switching efficiency we would like to have fast switching. If our switch has a rise time of 100 ns, or an equivalent frequency ($1/\pi t_r$) of about 3 MHz, we have no need to pass frequency components significantly above 3 MHz. If you select a filter cutoff of 10 MHz, there will be a barely perceptible change in rise time, but with huge benefits in high frequency control. In fact, such practice does not adversely impact the energy in the switch itself. Thus, there is no excuse for having a problem at frequencies more than 30 MHz. Nevertheless, we find switchers generating interference above 200 MHz. The key is to recognize that high frequency filtering can be applied without tangibly impacting the actual function of the converter itself.

In our experience such filtering must be applied as close to the switch itself as possible. One solution is to insert a ferrite in series with the switch, as shown in Figure 7-5. If you move the filter away from the switch, you open yourself to a large number of paths that must be pinched off.

Another issue is the current spike drawn during switching. Such a spike often occurs with push-pull designs where both switches may be on for a brief time, and sometimes occurs in other designs as well. Such current spikes are rarely appropriate in the converter design, and should be eliminated (they stress the components as well as cause interference). While it may be possible to filter out these spikes, the resulting degradation in rise time is almost always unacceptable. Accordingly, it is best to seek designs that inherently prevent such currents.

Snubbers

The abrupt switching transitions involving energy storage devices also often result in current or voltage spikes far in excess of that needed for operation. The spikes not only contribute to high frequency interference, but also stress the components. These should be avoided by employing snubbers to arrest the spikes. Figure 7-6 shows voltage and current snubbers at the switch. For those who are already employing such techniques, note that

Figure 7-5. Transient Suppression in Switch

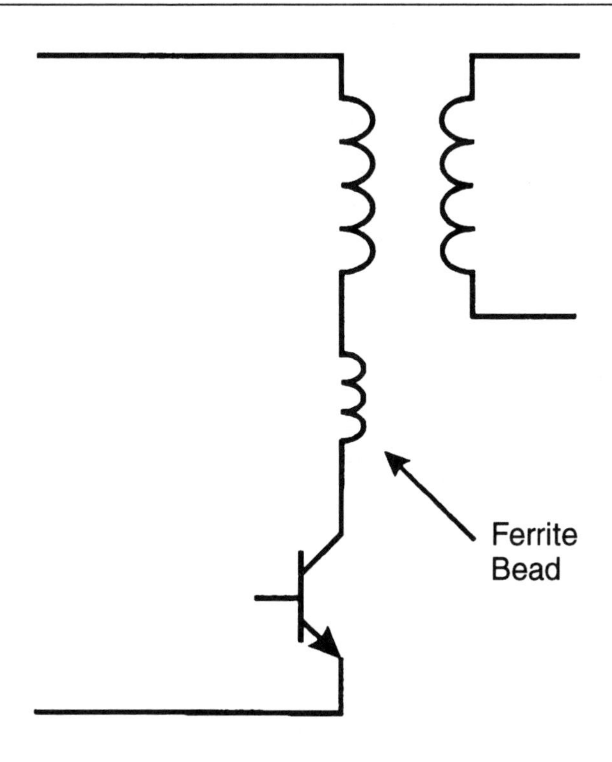

interference controls usually impose more severe limits on spikes than internal components require. So take a closer look at your existing snubbers and ensure that the highest frequency ringing components are adequately suppressed. You may not be able to detect such potential problems with your oscilloscope, especially if the bandwidth is low, but a spectrum analyzer will quickly identify these problems.

Supplemental Regulators

When there is a need for regulators separate from the power supply, several considerations are relevant. If there is a need for a low noise supply, then it is appropriate to select a linear regulator, otherwise a switching regulator may be employed. In both cases, however, it is important to provide local filtering, serving the purpose of controlling emissions and to block external interference. We recommend

Figure 7-6. Snubbers as Spike Suppressors

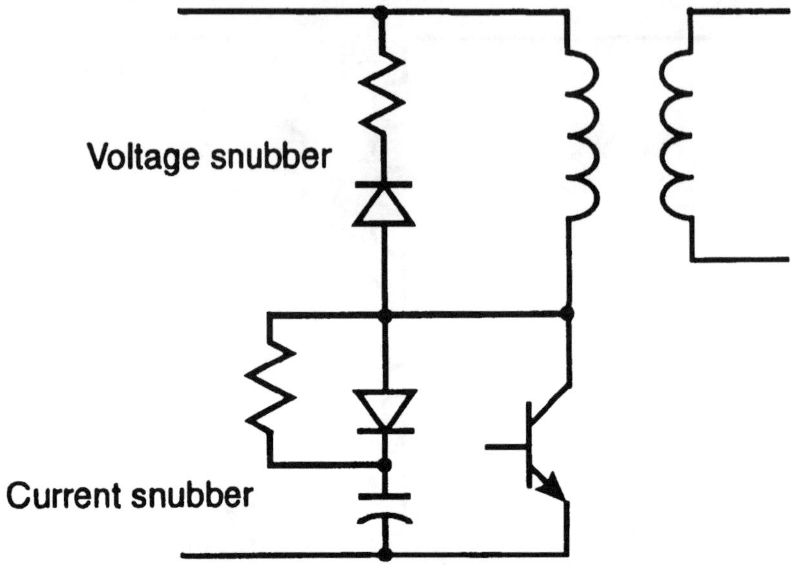

Voltage snubber

Current snubber

providing local filtering such as shown in Figure 7-7. Regulator technologies do vary; it is not possible to give a categorical solution to fill all cases, but it is usually permissible to insert a small capacitor in parallel with both the input and output. The common leg may or may not be at ground potential–the important thing is to ensure that the impedance path between the common leg and both the input and output is very low for the intended frequency. For switching technology this filter effectively confines the high frequency components to the immediate proximity of the regulator. For linear technology capacitors effectively block external interference from passing through the regulator to attack load electronics. For both technologies capacitors effectively keep the external interference from attacking the regulator feedback path. You may choose to supplement capacitors with ferrite beads to provide additional isolation in critical cases. If oscillation occurs, you may need to insert small (0.1 Ω) resistors.

IMMUNITY PROTECTION

Most attention to the power supply is typically centered on emissions from switching devices, but there is also a need to allow for

Figure 7-7. Regulator Filtering

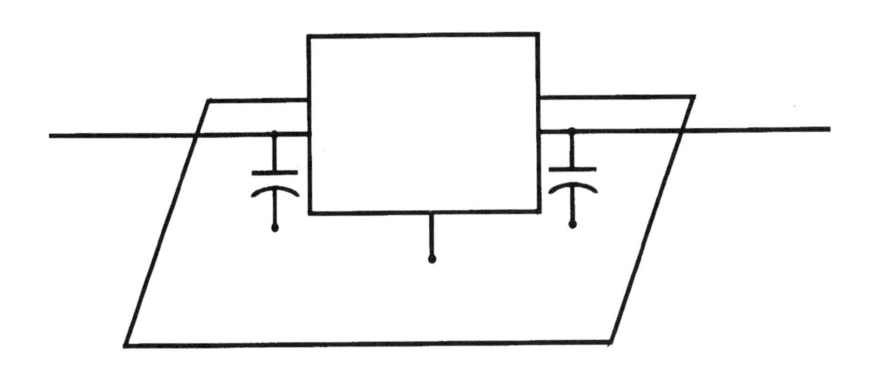

immunity to external disturbances, which may either attack the power supply itself or flow through the power supply to attack internal electronics. The problem boils down to two basic situations. First, the power supply itself is subjected to the interference that may overstress and damage the components or that may fool the regulator feedback circuit, resulting in erroneous output voltages. Second, the power supply filters that are primarily designed to control ripple are usually not capable of coping with high frequency components, so the interference simply flows through the supply.

A number of scenarios can cause trouble, and they are generally covered in the IEC 1000 documents. RFI (especially that in the hundreds of MHz) is often inadequately filtered, and flows through the power supply to attack internal circuits or attacks the regulator feedback path, giving erroneous output voltages. Most often, this is a result of the power feed wires intercepting RFI and transmitting it directly to the power supply. ESD is a high frequency, high current threat that, in the worst case, is of sufficient amplitude to destroy components directly in the path. Such a condition usually involves direct injection of ESD currents to a metallic conductor going to the supply. While this is not real common in medical electronics, it does occasionally occur, usually because one of the wires going to the power supply is either exposed (as might be the case with the ground wire) or insufficiently recessed, allowing arcing to an apparently inaccessible conductor, perhaps at an appliance plug. More often, the path is indirect: capacitively or inductively coupling to the equipment. In such a case the transient may either pass through the power supply, or it may attack the regulator feedback path. Depending on the design, the regulator path may be high impedance and the effects from the transient may last much longer than the actual transient itself.

Voltage supervisor devices, intended to provide a clean power-up, have become one of our most common problems. Power supply regulators are tricked into a momentary DC sag, which the supervisor interprets as failing power, shutting down the system and then resetting. If you are using such a supervisor, take extreme care to keep all interference away from the power supply regulator and the supervisor itself.

Power line transients include EFTs and surges. The EFT is a high frequency burst with peak amplitudes of 2 kV, generally arising from interrupting an inductive load nearby. The frequency components are much higher than can be handled by low frequency filters, and the amplitude is much higher than can be handled by a mild high frequency filter. After all, a 2 kV burst will still leave a 2 V burst after passing through a 60 dB filter, usually enough to upset digital electronics.

A surge is a lower frequency pulse, but has much higher energy, typified by a lightning strike. Such a surge may well be destructive to components, and may need transient protection.

Transient Suppressers for Immunity

Whichever case occurs, there are two basic methods of handling external interference–filtering or protecting with transient suppressers. Transient suppressers divert currents when the voltage rises above a certain level. The decision to use transient suppressers depends on the threat and the application. Where transient protection cannot be used, the burden must be placed at the filter, which now has the additional task of surviving a potentially large transient.

Two technologies for transient suppression were described in chapter 4. Clamps (large geometry Zener diodes and MOVs) are characterized by fast action and low power. Crowbars (arc devices) are characterized by slow action and high power. They can be combined into a single protector, as shown in Figure 7-8. The leading edge of the impulse is shunted by the MOV or Zener, which holds the voltage level down until the clamp has time to fire. Series inductors are needed to limit the current to the clamp, as the source impedance of the interference can be quite low.

Generally, for adequate transient protection, line to ground and neutral to ground suppression is needed, and this need has resulted in a conflict between transient protection and safety. In most cases the device will be subjected to a hipot test of 1500 V; therefore, the

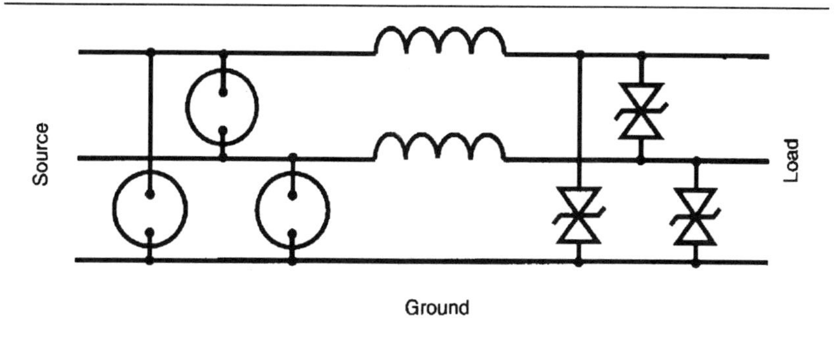

Figure 7-8. Transient Suppression

transient protector threshold needs to be above that, rendering any transient protection of limited value. At the present time line to ground transient protection should not be considered in devices, although such transient protection is available for permanent installation.

FILTERS

Filtering may be called upon to suppress emissions from the equipment and to suppress external interference. Filtering will normally include both DM and CM filtering. A common commercial filter is shown in Figure 7-9, which includes both CM and DM filtering. Common mode noise has previously been cited as difficult to handle with patient-connected medical equipment, due to leakage current limitations that sharply restrict the capacitance from power to ground. Filters designed for medical devices omit the CM capacitors (called "y-caps").

Fundamentally, effective filtering is accomplished by a combination of series inductors and shunt capacitance. Generally, for any given set of conditions (source and load impedance) there will be an optimum filter topology of capacitors and inductors, but there are trade-offs that can be made.

Where leakage current is limited, CM shunt capacitance is limited, and the only option is to increase series inductance. Even moderate filtering needs can place a significant burden on inductors; we often find that inductors take up more room than available. Even if there is room for inductors, they will be placed to provide a high impedance facing the source of interference. High impedances are hard to maintain at high frequencies–stray capacitance in the

Figure 7-9. Power Line Filter

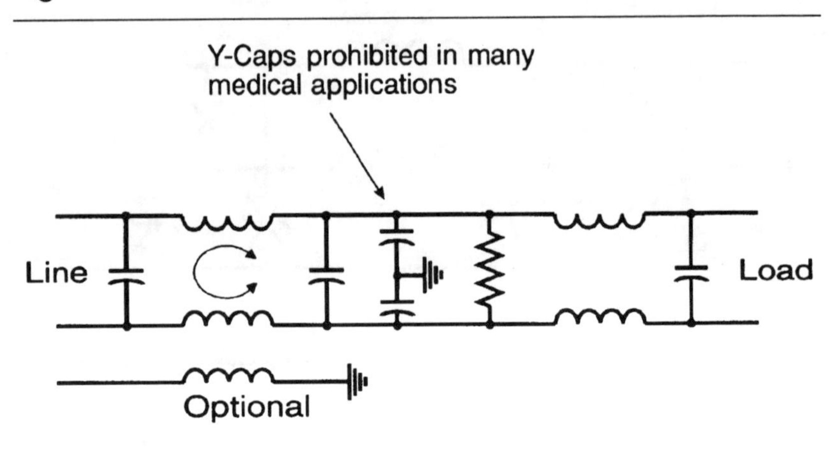

windings and in nearby circuit and structural elements provide sneak paths around the inductance. This is why we emphasize minimizing these CM currents before they start.

Differential mode power supply noise dominates up to about 1 MHz. Above 1 MHz, CM effects become dominant, as coupling paths become more efficient.

Inductive elements must be placed so as to minimize coupling. This can be accomplished by orthogonal mounting of adjacent components. Open loop inductors are a particular problem, as the magnetic field extends well beyond the boundaries of the inductor. Two inductors, mounted side by side, form a loosely coupled transformer, reducing CM suppression of inductors.

Filter Design

Designing a power line filter is no longer a simple task. There are numerous constraints that must be satisfied. We would prefer that the filter be designed by the power supply designer, as the filter can adversely affect the operation of the power supply. Recognizing that a regulated power supply is essentially a negative resistance device, you need to be particularly careful that you do not create instabilities causing the power supply to become an oscillator. This is no small risk when employing reactive elements immediately facing the power supply.

There are other constraints, too. You need to ensure that your filter does not simply starve the power supply. You need to ensure that you have adequately provided for both CM and DM suppression. You need to ensure that the filter is small enough to fit into the space available. You need to ensure that the filter operates as intended when mounted in the product. Finally, you would like the filter to be cost-effective.

Your placement of the filter elements (T, L, or π) depends on the impedance of the source and the load. In some cases the selection is critical. A prime example is the output to a stepper motor or a variable frequency drive. In these cases the output impedance of the driver is very low. It is inappropriate to put a shunt capacitor across the output of these drivers–the motor will not work. The correct answer is to start with a series inductor. Filter houses have filters designed expressly for this purpose. We generally advise using a commercial filter house wherever possible–they know most of the pitfalls. If not, here are some pointers to get you started in the right direction.

The approach to filter design is to design the CM and the DM filter separately, then combine them into one. While it appears from the schematic that the CM and DM aspects are thoroughly merged, a closer look at the component values will reveal that the DM filter is aimed at a lower frequency than the CM filter. This is because of some fundamental differences between the two modes. Let us take a look at these.

First, DM tends to dominate below 1 MHz, whereas CM tends to dominate above 1 MHz. The reason is simple once you stop to think about it: Common mode originates from coupling paths, which are more efficient at higher frequencies. Differential mode interference tends to stay on the intended conductor path.

Second, the demands on inductors are much different. Common mode chokes carry little or no net current, as DM currents cancel. Common mode chokes can use high permeability core materials, as there is no risk of saturation. The higher frequencies encountered in CM require the use of high frequency core materials, generally a ferrite. Differential mode requires individual chokes, which must carry all the current to the power supply; thus, core saturation is a concern and low permeability materials are used. As the frequency is lower, lower frequency core materials are usable, such as powdered iron or molyperm.

Third, the impedance of the source and loads are generally different for CM and for DM. At the power supply the DM source tends to be inductive; therefore, the first element in the DM filter would normally be a capacitor. The CM source tends to be capacitive; therefore, the first element in the CM filter would normally be an inductor. At the power line side a similar situation exists: the DM tends to be inductive and the CM tends to be capacitive. Thus, you would optimally have the DM element closest to the line be a capacitor and the CM element closest to the line be an inductor.

Finally, the capacitor needs to reflect the different voltages expected from line to ground and from line to line. The line to ground capacitor must usually pass the 1500 V hipot test, where line to line capacitors can work with a lower voltage.

Differential Mode Filters

Start by making some basic assumptions. First, we would like the filter to have a minimum impact on the power supply; therefore, we start by assuming that we can insert a 1 percent influence without adversely affecting the power supply. Thus, we would like to have series inductance and shunt capacitance exert no more than a 1 percent influence on the power supply load. So start with impedance of the inductor at 60 Hz being $< 0.01Z$ (load), and impedance of the capacitor at 60 Hz being $> 100Z$ (load). Generally, it is best not to put a capacitor directly across from power supply inputs, as oscillations may occur. If oscillations do occur, insert a small value damping resistor in series.

The inductor core must be sized to avoid saturation. Typically, use low permeabilities (of the order of 100), such as found in powdered iron or molyperm. Check to ensure that the core resonance is comfortably above 1 MHz. If resonance is too low, split the inductor into two or smaller inductors in series.

Common Mode Filters

Generally, start with the maximum allowed leakage capacitance to ground, then supplement with common mode chokes until the desired attenuation is reached. If you are building diagnostic equipment, then you may be allowed several nF. If you are building patient-connected equipment, then you may have only 100 pF or even less. Divide the allowed capacitance equally among the number of lines (line and neutral, three phase, etc.). Common mode chokes should use the formulation suitable for the frequency range of interest. The most effective formulations for CM filtering have a peak impedance in the 10–20 MHz range. You are likely to need a lot of series impedance, so shoot for an inductance of at least 5 mH, which is easily achieved using high permeability ferrites. Saturation of CM chokes is not an issue, but again, look to keep the resonant frequency high enough that the choke is still effective at the maximum frequency of interest. For immunity threats this frequency may be 1 GHz.

A Warning When Selecting a Catalog Filter

Filter attenuation is specified using a 50 Ω source and a 50 Ω load. But, in practice, neither the source nor the load are ever at these impedances. Power line impedances at low frequencies are very low, but as frequencies increase, the source impedance swings widely from a very low impedance (fractions of an ohm) to about 100 Ω, as resonances are encountered. Similarly, the power supply load is usually a full wave bridge, which is either very high or very low depending on whether the diodes are in conduction or not. Thus, you may be certain that you will never reach the attenuation in your application that is specified in the catalog. You will be much closer to the mark if you subtract 20–40 dB from the published numbers.

It may be argued that the standard test allows for relative comparison, but we find that not to be true. If the topology is the same, you will come fairly close, but the correlation between single stage and multistage filters is not very good.

How Much Attenuation Do You Need?

We have given the guidelines, now let us try to understand how much attenuation is needed. First, we need to be reminded that the published attenuation is not a reliable indicator of attenuation–it is specified at 50 ohms in and 50 ohms out. We need to be realistic and determine just what impedances are suitable.

If your concern is emissions, then your equipment is tested at 50 Ω line impedance. The load, of course, is the bridge rectifier. Your best assumption is that the impedance is either 0.1 Ω or 100 Ω. It is not possible to tell how much attenuation you need without knowing what the emission amplitude is, and this is a strong function of the power supply design. The levels can be easily tested, however. Once you have the amplitude and frequency spectrum, you can calculate the amount of reduction you need.

If your concern is immunity, then your equipment will be facing a known source, but you will be left with a unknown susceptibility at the load. The line impedance will similarly vary between 0.1 and 100 Ω, and an approximation of effects can be made.

How much of a 2 kV spike can you allow to pass the filter?
If it is 2 V, then you need 60 dB attenuation (1000×).

Generally, though, you will be faced with the prospect of designing or selecting a filter with no prior knowledge of what filter is needed, so you need a starting point. We like to start with a filter attenuation need of 40–60 dB at the lowest frequency of interest. As frequencies increase, attenuation needs usually diminish a bit. If you are buying your filter, you should derate the specifications by at least 20 dB. If you are designing your own filter, use the source and load impedances mentioned above, and compute the attenuation at the lowest frequency of interest. For DM, this will be 150–450 kHz. For CM, the lowest frequency will typically be 1 MHz. You will also need to ensure that your filter elements are effective at the highest frequencies of interest, too. Remember that filter elements are effective for only two or three decades of frequency, as the parasitic factors in the component limit the upper frequency range. So you will need to look at higher frequencies for radiated interference out of the power cord and for immunity issues.

Component Selection and Placement

Filters are intended to block electromagnetic energy, so you would expect to find considerable electric and magnetic fields in the components and in the immediate proximity. It is important that the filter elements be selected or designed to minimize self-parasitics–the capacitors should have minimum ESR (effective series impedance) and the inductors should be wound to minimize capacitance in the windings and also to minimize the spread of the magnetic field outside the core boundary.

Capacitors should be selected to be effective in the frequency range of interest. High frequency performance is not as important with DM interference as it is for CM interference. But CM interference problems can occur up to 1 GHz, and high frequency performance is imperative. In either case the leads should be kept short to minimize lead inductance.

Inductors should be wound to minimize interwinding capacitance. Good and poor winding practices are shown in Figure 7-10. In case A the windings provide for large potential differences between some adjacent windings, multiplying the effective capacitance. Case B is wound to minimize the maximum voltage between windings. RF chokes are commonly wound in groups of three, to minimize capacitance, but this practice involves an open flux path inductor, the use of which is discouraged.

Inductors should be a closed flux path if possible. For large cores this means a toroid or E-core (pot cores are better yet). If core gaps

Figure 7-10. Inductor Winding Practices

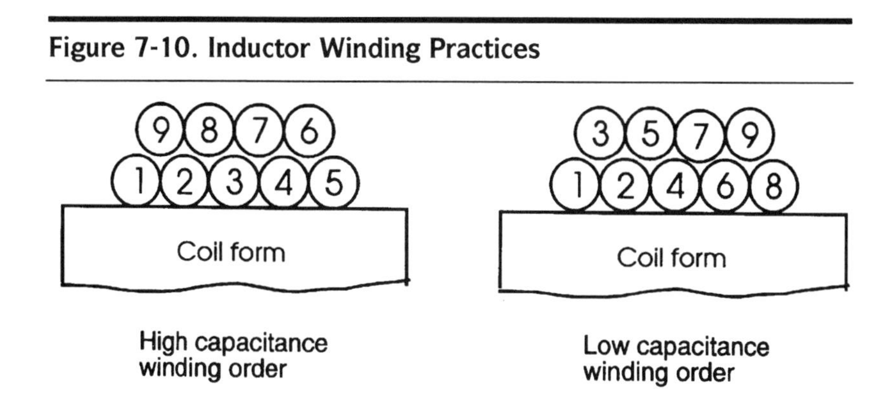

are required, they should be spread equally around the core. If an E-core must be gapped, gap the center leg.

Filter elements must be mounted so as to minimize both capacitive and inductive coupling. Cores should be mounted orthogonally to minimize coupling, especially where open flux path inductors are used, but the practice is useful even for toroids. Keep the noisy end of the inductor clear of heat sinks, circuit board traces, or structural members–electric field coupling is a surprisingly significant factor in inductors, especially where high value inductors are needed.

Filter Mounting

Filter performance, especially at the higher frequencies of interest, is highly dependent on mounting. In fact, we see more filter failures due to mounting than to inadequacy of the filter itself. The factors that diminish the effectiveness of the filter are shown in Figure 7-11. A high impedance mount (either due to an inductive pigtail or to a high resistance bond), converts the filter capacitors into paths bypassing the inductor. Stray capacitance provides a high frequency shunt around the filter. The leads after the filter are exposed to a noisy field. All three of these deficiencies are eliminated in one fell swoop by a firmly bonded bulkhead mount as shown. This explains the popularity of the bulkhead mount. If it is not possible to mount to the bulkhead, then the filter should be bonded directly to the chassis in close proximity to the power entry point.

Figure 7-11. Filter Mounting

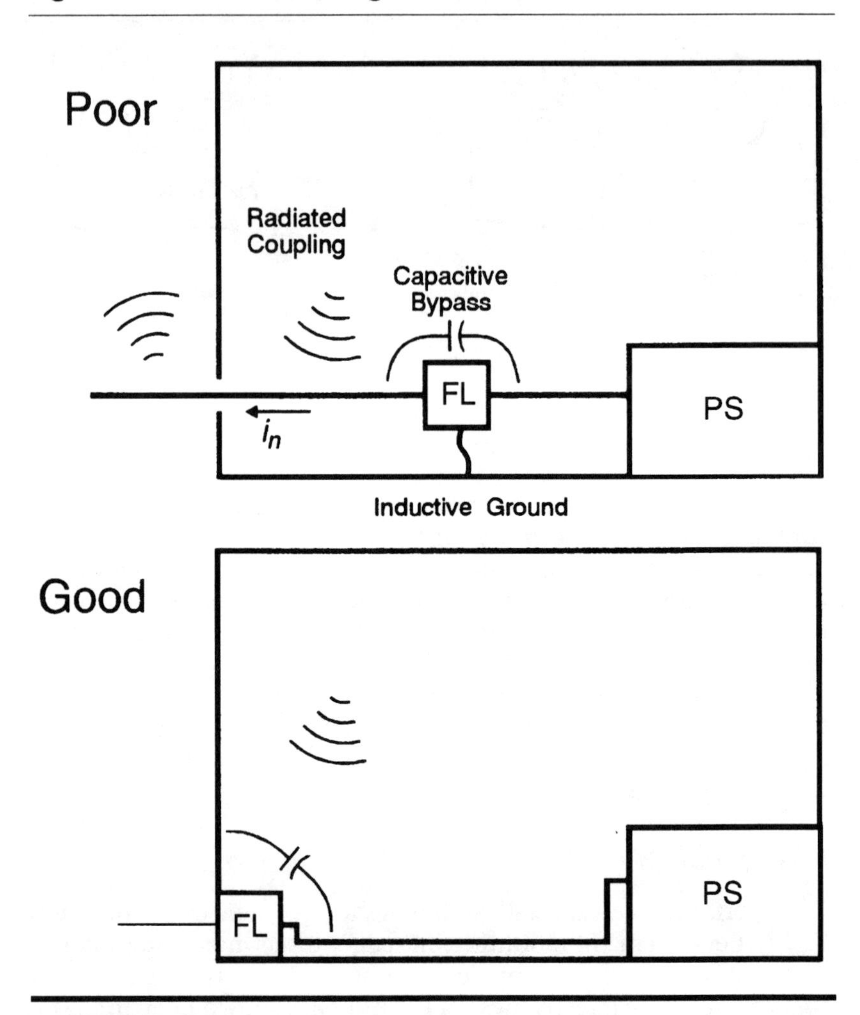

SUMMARY

Power supplies and filtering needs can be summarized into a few basic rules:

- Rectifier bridges and switching converters are major sources of interference.
- Interference originates as DM, but coupling paths are primarily CM.

Figure 7-12. Commercial Filter Designed for Bulkhead Mount (Photo Couresy of Corcom)

- Differential mode interference dominates below 1 MHz, whereas CM interference dominates above 1 MHz.
- Differential mode interference is best contained immediately at the source by filtering and circuit design.
- Common mode interference is best contained by component placement to minimize inductive and capacitive coupling.

- Filter design needs to be compatible with power supply design.
- Filter design needs to account for DM and CM interference. Common mode interference is difficult to achieve where leakage currents are limited.
- Filter elements need to be sized to be effective at the desired frequency range.
- Filter elements need to be placed to minimize adjacent component coupling.

8

INTERCONNECT AND MOUNTING

The internal construction and even the selection of wires within a cable can affect the EMI performance of a cable. This applies to both internal and external cables used in a system.

This chapter addresses the aspect of transferring data, information, or power from one part of your equipment to another (chapter 10 discusses external cable issues). These issues are of particular interest to diagnostic systems, where a number of modules may be mounted in one system; however, the principles are relevant to any system that involves cabling of one type or another, even if it is only a short ribbon from one board to another or a single cable to the outside world.

The issue is that of transferring data from one module to another with minimum signal degradation, which is relevant to self-compatibility, immunity, and emissions. As it is difficult to adequately address these issues without a sound enclosure, we will also discuss enclosure design.

CABLES AS TRANSMISSION LINES

We have already seen cables as antennas, due to their physical dimensions. For the same reason cables also act as transmission lines, and are subject to reflection and crosstalk problems. Since these problems are inside the system, they are not always considered true EMI problems, but rather "signal integrity" problems. We do consider them EMI problems, however, since they can and do adversely affect system performance.

Reflections and Terminations

Long lines must be treated as distributed, since there is a delay from the time the signal is injected into the line to the time it is received at the end of the line. Thus, the signal propagates along the line without regard to the eventual termination of the line. When the signal reaches the far end, it will be either absorbed or reflected, depending on the termination. These reflections can cause circuit upsets, and can even contribute to increased radiation of the lines.

Transmission lines can be represented by small lumped impedances, consisting of series inductance and shunt capacitance. This yields a characteristic impedance for the line, given by

$$Z_0 = \sqrt{\frac{L}{C}}$$

where Z_0 is the characteristic impedance in ohms, L is the inductance per unit length, and C is the shunt capacitance per unit length. Z_0 is the impedance first seen by a source, before any reflections occur that provide information on the actual termination of the line.

Reflections occur whenever the line is terminated in an impedance other than Z_0 Figure 8-1 shown examples of reflections due to both open and short circuit termination. For digital signals these reflections become an issue when the rise time approaches the round trip propagation delay along the line. The resulting reflection will add or subtract, depending on the load and source impedance, resulting in overshoot or undershoot. The most common situation in digital interfaces is a low impedance source and a high

Figure 8-1. Transmission Line Reflections

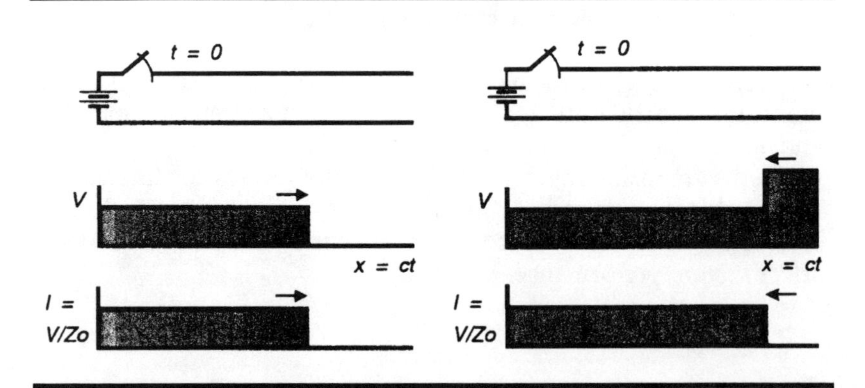

Figure 8-2. Waveform Ringing Due to Impedance Mismatch

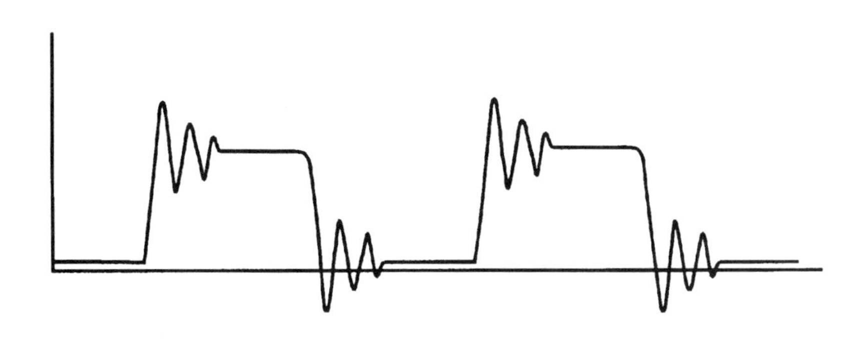

impedance load, which results in the "ringing waveform" shown in Figure 8-2.

Table 8-1 gives some recommended lengths vs. edge rates for determining when terminations are needed. These are based on the round trip transit time equal to the rise time, and assume the speed of light for "loose" cables, and about 50 percent of the speed of light for circuit boards and ribbon cables. It is clear that high speed signals typically need terminations if the cables are more than a few meters in length. Alternately, the rise times can be slowed with filters for interfaces that do not require high speed data transfer. Actual termination methods are addressed in chapter 5.

Crosstalk: The Hidden Transmission Line

Crosstalk is a very common problem in cables, and arises from near-field electric and magnetic field coupling between adjacent lines in the cable. In a sense this near-field coupling forms a parasitic transmission line between unintended conductors. As the frequencies (or

Table 8-1. Length vs. Edge Rate for Determining Termination

Edge Rate	Distance in Loose Cable	Distance in Ribbon Cable
1 ns	15 cm	7.5 cm
10 ns	1.5 m	75 cm
100 ns	15 m	7.5 m
1 μs	150 m	75 m

edge rates) increase, more and more energy is transferred to an unintended receptor circuit. In high speed digital interfaces this can result in "spikes" that jam the adjacent line. But even in low frequency analog circuits, small amounts of crosstalk can jam very sensitive receptor circuits.

Figure 8-3 shows a cable crosstalk model. Note the two coupling modes–capacitive and inductive. Both modes are present, although capacitive coupling predominates in high impedance circuits, and inductive coupling predominates in low impedance circuits. As a result, the capacitive mode is usually the culprit in digital and high impedance analog interfaces, while the inductive mode is usually the culprit in power interfaces and low impedance analog interfaces.

Although predicting the exact values of crosstalk can be complex, fixing or preventing the problems are relatively simple. Here are some recommendations:

To reduce *capacitive* coupling:

1. Increase wire separation and/or parallel run (reduces common capacitance).
2. Decrease separation between signal and associated return lines (increases shunt capacitance).
3. Slow down edge rates or frequency (increases capacitive coupling).
4. *Decrease* circuit impedances (may increase inductive crosstalk).
5. Add a shield or guard trace (interrupts capacitive path).

To reduce *inductive* coupling:

Figure 8-3. Cable Crosstalk Model

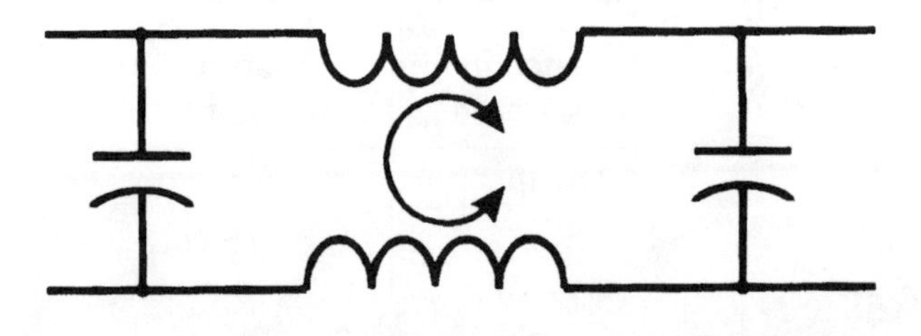

1. Increase wire separation and/or parallel run (decreases mutual inductance).
2. Decrease separation between signal and associated return (decreases mutual inductance).
3. Slow down edge rates or frequency (decreases mutual inductance coupling).
4. *Increase* circuit impedances (may increase capacitive crosstalk).
5. Twist wire pairs (provides some cancellation of the inductive path).

Note that for both modes the first three recommendations work equally well. Only the fourth recommendation, changing circuit impedances, are in opposition. The fifth recommendations are specific to the particular mode, but do not conflict.

Since crosstalk is similar to transmission line problems, the same guidelines on dimensions apply. That is, crosstalk is most likely a problem at distances where transmission line effects prevail.

Like many EMI issues, crosstalk defies simple analysis. Popular models assume lumped circuit analysis, which has been shown to produce significant errors. Nevertheless, it serves for discussion purposes. Distributed analysis or empirical analysis will give better results, but computer analysis is often necessary.

We find, however, that most crosstalk problems occur, not because of misanalysis, but because the designer never considered crosstalk in the first place. In fact, we handle most crosstalk issues by inspection, using a simple criteria.

Interference always starts as differential mode (DM), but coupling paths are always common mode (CM). Differential mode is generated between two terminals of a device and generally propagates normally as a noise/return path. As long as the interference stays between these wires, the interference remains DM. But coupling paths from a noise source are largely CM. Either inductive or capacitive coupling from a wire of a circuit member to an adjacent cable bundle is largely common to all wires in the bundle. (This commonality is far from 100 percent, however.)

Thus, we can make several inferences about CM and DM. First, the closer to the source of interference, the higher the percentage of DM exists. Immediately at the source, interference is almost entirely DM. As you get farther away from the source, you have more opportunity to couple, therefore, you increase the percentage of CM.

The second is that since CM is always arrived at by field coupling paths, and these paths are more efficient at higher frequencies, we expect CM to dominate at higher frequencies and DM to dominate at lower frequencies.

As a crude rule of thumb, crosstalk starts to become a significant issue when the entire wavefront is impressed along the parallel coupling path. Thus, if the rise time of a signal were 10 ns, and the speed of light down the wire is 1 ns/ft, then we would expect crosstalk to be a problem for cable lengths of about 10 feet or longer. Obviously, this is an extremely simple case, and actual cases will vary widely, depending on whether a ground plane is present and signal sensitivity.

COMMON MODE AND DIFFERENTIAL MODE ON CABLES

Now that we have determined that cables are often unintended antennas, we need to look at a concept crucial to understanding cables: *CM* vs. *DM* EMI. Common mode refers to EMI currents that flow in the same direction on all conductors in the cable, while DM refers to EMI currents that flow in both directions. This is illustrated in Figure 8-4.

Note that CM currents must return through some path other than the cable. This may be another cable, the "ground," or even some "sneak path." Since the CM current flows along the length of the cable, it is often called the "longitudinal" mode. Differential mode currents, on the other hand, flow in both directions, so the current return is in the cable itself. Since this is the direction that

Figure 8-4. CM and DM Currents

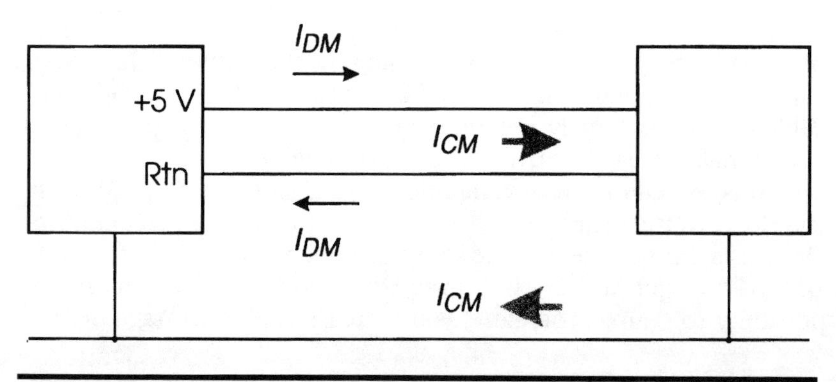

normal power and signal currents flow, this mode is often called the "normal" mode.

The two modes are illustrated in Figure 8-5. Note that both modes can be present at the same time, and that both power and signal cables are affected. In fact, power cables have two common mode cases—a general case where noise currents flow in CM on all lines, and a special case where noise currents flow CM on the power and neutral, yet return via the safety ground (green wire).

These mode distinctions are very important, for they often require different EMI strategies and fixes. For example, line-to-line capacitors will filter DM noise, but will have no effect on CM noise, since the different lines are already moving in the same direction. In this case line-to-ground capacitors are needed to filter CM currents. As we will see later in this chapter, wire twisting is very effective on DM EMI, but has no effect on CM cable currents. As another example, a ferrite clamped over a cable will suppress CM currents, but will have no effect on DM currents.

These modes also gives us some clues as to the origins of EMI currents. For example, DM currents suggest that conductors are connected to the source circuit, while CM currents suggest that EMI

Figure 8-5. CM and DM on Power Feed

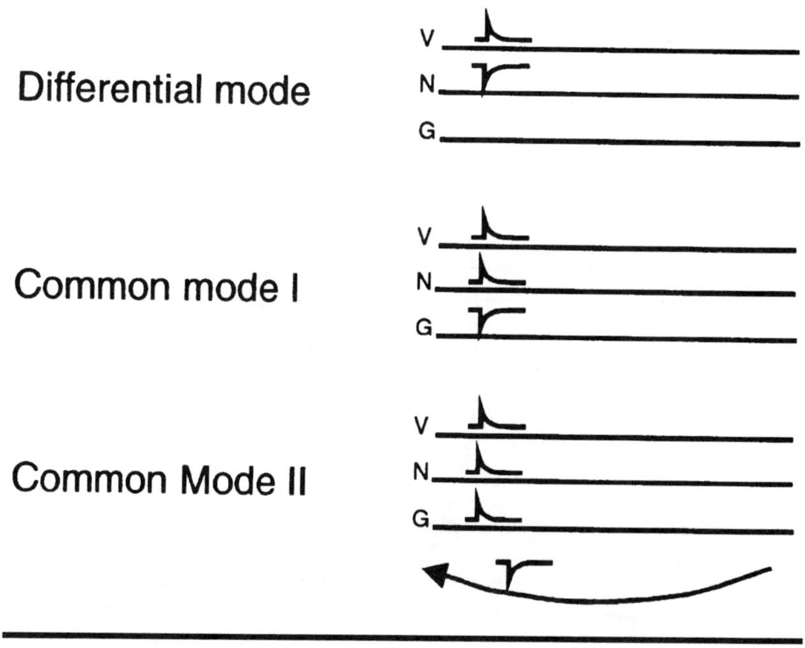

Differential mode
V
N
G

Common mode I
V
N
G

Common Mode II
V
N
G

energy has been coupled to the cable through radiation, crosstalk, or common ground impedances. The special power line case discussed above (safety ground return path) usually indicates parasitic coupling in the power supply, which was discussed in chapter 7 on power supplies.

The coupling modes can be determined experimentally with a current probe, as shown in Figure 8-6. A current probe is nothing more than a loosely coupled transformer that uses a single turn of the cable as the "primary" winding. The output voltage is proportional to the input current–thus, the "current probe." Current probes come in a wide range of frequencies and current levels. For EMI work you want a cable probe that functions up to several hundred MHz if you want to be able to detect EMI currents on cables.

Figure 8-7 shows how to determine the current direction. For simplicity this is demonstrated with a power cable, but the

Figure 8-6. Current Probe

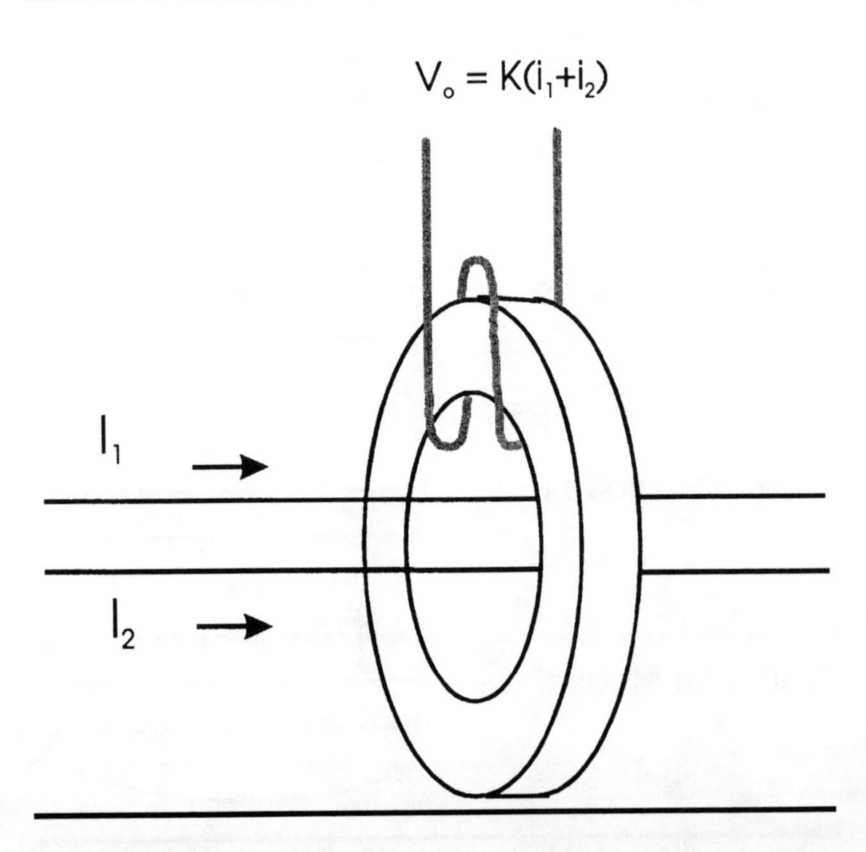

Figure 8-7. Using Current Probe to Determine DM and CM Currents

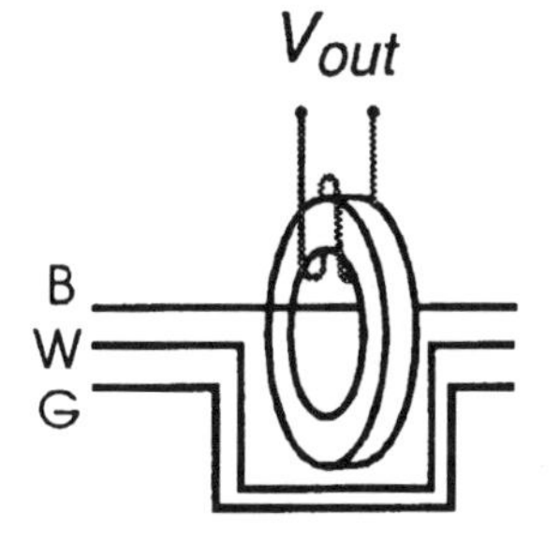

Clamp probe over each wire. Voltage out is proportional to current on each wire. Current may be DM or CM.

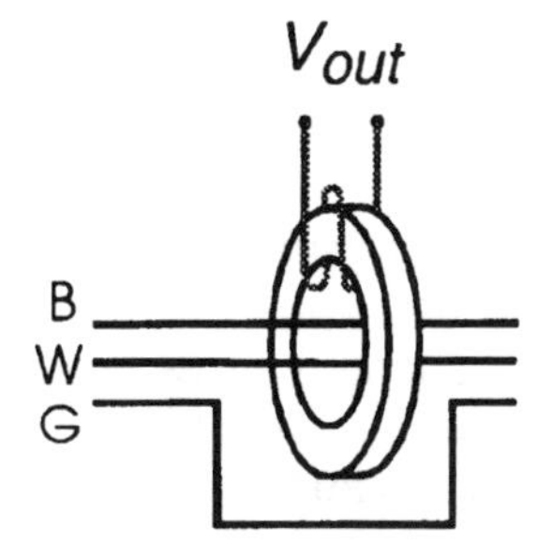

Clamp probe over both B and W wires. If V = 0, currents on B and W are DM. If V is not zero, CM currents exist.

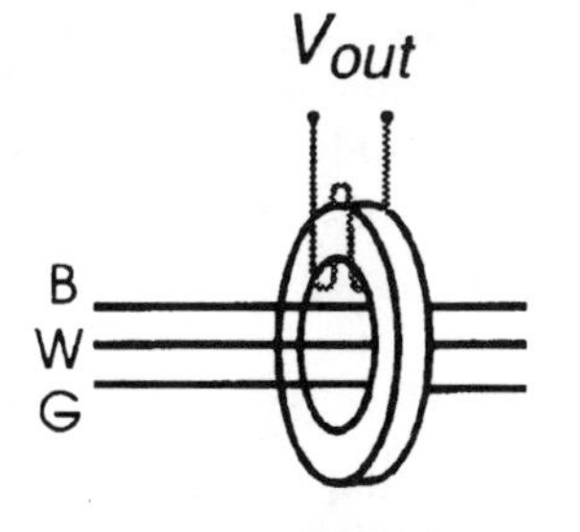

Clamp probe over entire bundle. If V = 0, currents are CM with respect to G. If V not zero, CM is on all three wires.

technique works equally well with signal cables. First, clamp the probe over individual lines. If current flow is indicated, it can be either CM or DM. Next, clamp the probe over the entire cable. If you still show current flow, then you have CM current. If, on the other hand, you show no net current flow, the current flow is DM, since you have "equal and opposite" currents now flowing through the core of the current probe. You can even try this with individual wire pairs, to determine specific DM current paths. The current probe is a very useful tool. It is often referred to as the EMI engineer's stethoscope for cable problems.

The insidious and frustrating part of CM currents is that they are very difficult to predict. Furthermore, they do not show up on any schematic. They are usually due to some parasitic effect, like common ground impedance or unexpected capacitive or inductive coupling. Furthermore, since the intended currents follow DM paths, some kind of DM to CM conversion must take place for CM currents to exist. Conversely, CM currents must be converted back to DM currents at the receptor circuit. If we can determine where these conversions occur, we can often cure the associated EMI problems. Unfortunately, these "conversion points" do not show up on schematics either. You need to understand the problem so you can find them.

Common mode currents on cables are a major cause of high frequency EMI problems, for both emissions and immunity. Most Federal Communications Commission (FCC)–radiated emission failures are due to CM currents on a signal or power cable, rather than DM currents. Most RF immunity failures are also caused by the CM currents resulting from cable antenna effects. Even low frequency problems are due to CM coupling, usually due to improper grounding.

Differential mode currents, however, can also cause EMI problems. Crosstalk within a cable is a DM problem. (Common mode crosstalk also can exist between cables.) When dealing with EMI problems on cables, you must assume both modes are present. Chapter 5 discussed how serious CM cable coupling can be.

CABLE LAYOUT AND PINOUTS

A very common problem for digital cables is inadequate ground or signal returns. The first place to look at in any interconnect system is the cable pinouts. All too often, you have no choice over the pinouts—they were defined by others and you are stuck with them. But sometimes you have control over them and you should take advantage of this fact. Eventually, as others catch on, poor pinout design will diminish.

Figure 8-8(a) shows the situation in a ribbon cable, which suffers from a number of deficiencies. In an amazing number of cases, this same scenario exists: The designer starts the pin assignments at the top with pin number one and works on down. Data is pin 1-8, address is pin 9-16, and so on, until you get to the last signal that is at pin number 37. You select the next largest ribbon cable, which is 40 conductors. The remaining unused pins (38, 39, and 40) are assigned to ground. This results in a poor pin configuration, from a self-compatibility standpoint and from an immunity and emission standpoint:

- There are too many conductors for the number of grounds. Remember that ground noise is a signal to the receiving chip.
- Adjacent signals are fully exposed to adjacent line coupling, or crosstalk.
- The loop area formed by the ground path and the farthest conductors is huge, making good antennas for emissions or immunity.

A far better cable pinout assignment is to provide more ground pins and to distribute them throughout the cable. Figure 8-8(b) shows a more favorable example. Now the grounds are spread throughout the cable, minimizing loop areas, and minimizing both

Figure 8-8 Good and Poor Cable Pinouts

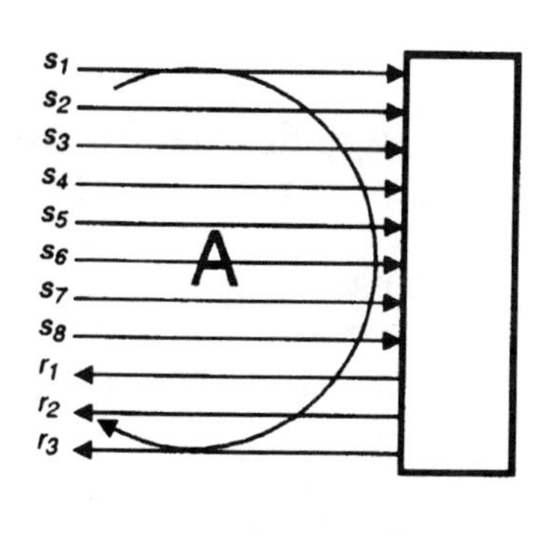

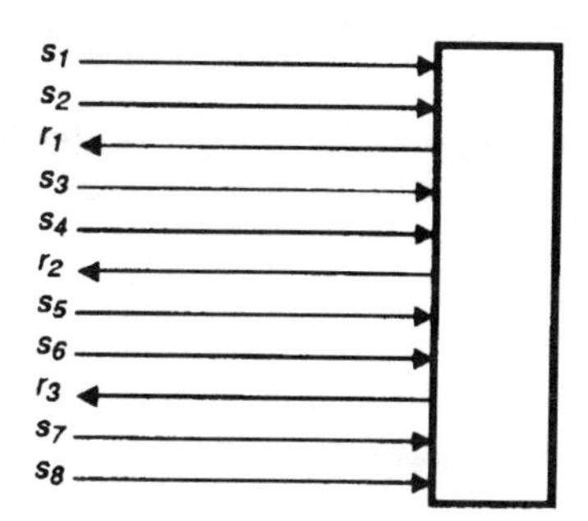

(a) Poor pinouts (b) Good pinouts

emissions and reception of external interference. In the illustration each signal line is close to a return path; thus, adjacent line crosstalk is minimized. Finally, there are only a few lines sharing a ground return, minimizing ground bounce.

As a rule of thumb, we recommend no more than five signal lines per return line for digital interfaces with rise times greater than 3 ns, with a dedicated return line for each signal line preferred. With rise times less than 3 ns, you need a dedicated return for each single line. These return lines also act as "guard traces" to minimize capacitive crosstalk.

The best, of course, is one return for each signal (better yet, twisted pairs), as would be needed in high speed circuits like ECL. The second best is one return for each signal pair (as per the illustration)–this provides a return adjacent to each signal line.

You may be able to extend the utility of your available grounds. If you are concerned with emissions, you can concentrate the grounds near high speed lines, especially the clock. If you are concerned with immunity, you can concentrate the grounds near the sensitive analog lines. If you are driving DC power down the line, and you decouple well at both ends, then the DC may be considered almost as good as a ground.

The key, then, is to fight for plenty of signal grounds. We have lost count of the number of times we have been unable to solve the problem until we got more signal grounds. Conversely, we have never had a ground bounce problem when we have had the recommended number of grounds. The worst case we saw was when there were 79 signal lines and only one return conductor–the case involved two 40-pin cables and only one had a return path.

SIGNAL BANDWIDTH

Along the same line remember that any energy that need not be transmitted can be filtered either at the driver, the receiver, or both. Generally, our rule is to transmit only the bandwidth needed to obtain adequate signal integrity at the receiver plus an operating margin. Anything more either leads to excessive emissions or crosstalk. By the same token the receiver should be filtered to minimize bandwidth as well.

ROUTING YOUR CABLE

Internal cables, whether signal or power feed, should be routed adjacent to a ground surface to minimize loop areas. This will minimize CM emissions, reduce cable-to-cable crosstalk, and enhance immunity. Bulk ferrites may be placed over the entire cable bundle to help reduce CM. Generally, such a patch is good for no more than 10 dB (a factor of three).

Where there are many cables, try to group similar types of cables together, then route all next to the ground. In particular, do not route noisy lines adjacent to sensitive lines. Tie wrapping is one of the principle causes of adverse crosstalk between cables.

It is important that the cable be routed along a continuous ground. Remember that CM currents on a cable will try to return along the ground immediately under the cable, if the cable is very close to ground. If the ground path is interrupted by a seam, the return current must go around, energizing the seam in the process–making a fine antenna (Figure 8-9).

In general, leaving cables of any type hanging in midair is an invitation to trouble. Each system will operate differently, depending on cable placement. In fact, you will drive yourself nuts trying to troubleshoot a system in which the cable positions are not controlled–each time you remove the cover, the cables move to a different location, giving different results.

Where intra- and intercable coupling cannot be reduced to acceptable levels, then the last resort is cable shielding. These techniques are discussed in chapter 10. Chapter 10 focuses on external cabling, but the principles apply to internal cabling as well.

DESIGNING AN ENCLOSURE

You cannot route your cables properly if the enclosure is poorly designed. So here are some guidelines, which take good grounding practices and good cable practices and puts them together into a single set of complementary rules.

Enclosure Ground

Start with a ground reference–a plane that is accessible to all modules. The ground should be a wide plane, but may be shaped to meet needs, including making a "T" or a bend. It may have holes for cable access. It should be continuously conductive (welded is excellent). Bolting together of painted or anodized metal members is not satisfactory. If mechanical splices are necessary, make the mating surfaces conductive and put fasteners at frequent intervals (every inch or so), especially where cables are routed across. Hanging a metal skin over a wire frame does not constitute a ground.

Figure 8-9. Cable Radiation Due to Routing Over Seam in Ground

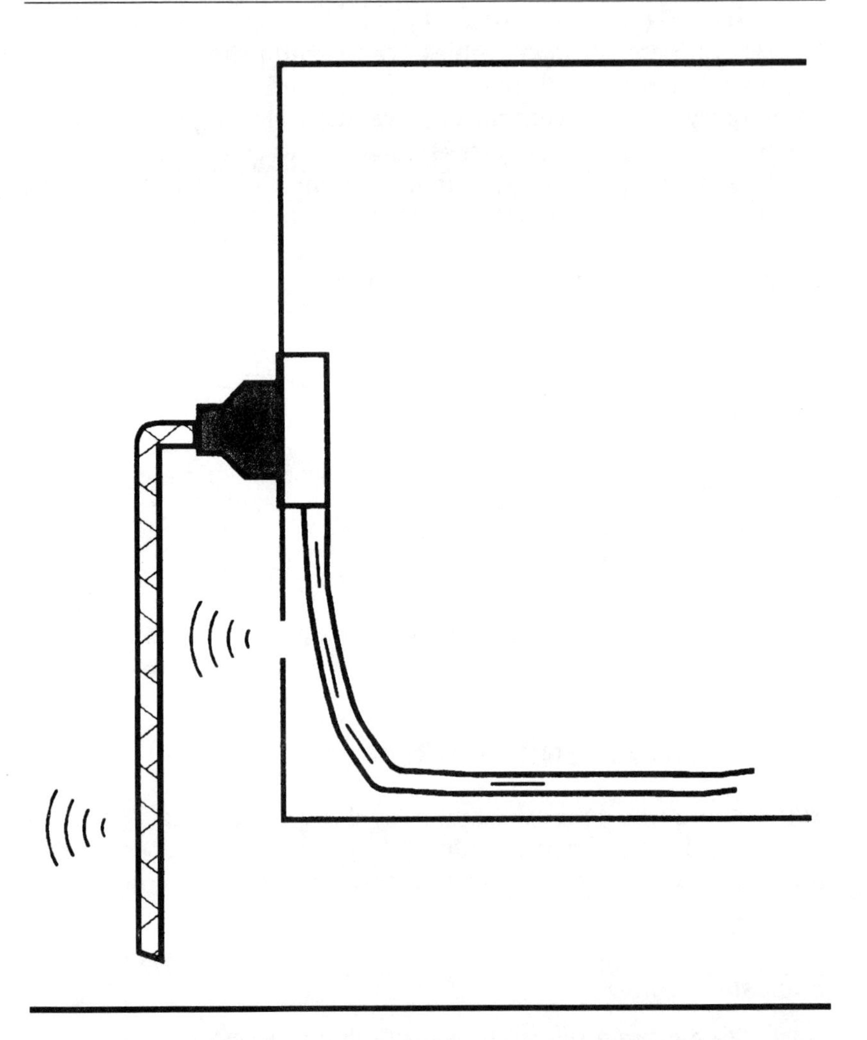

Purchased card cages are usually very poor–their ground is essentially nonexistent. If your card cage consists of a dozen pieces loosely fastened together, you do not have a ground.

Bond all modules to ground reference, preferably directly, or at worst, a short wide ground strap. Avoid daisy-chaining grounds; if a module is some distance from the ground plane, and you can considerably shorten the ground path by connecting to a module ground, then do it rather than run a long ground wire.

Mount all external power and signal cables to a single metal plate, preferably the ground plate, but if this is not possible, be sure the connector plate is very well grounded. This practice is to ensure that your equipment does not become an inadvertent victim of external ground currents (Figure 8-10).

Power Distribution

Mount EMI filters, transformers, and protection devices directly to the window immediately at the enclosure boundary. Distribute both AC and DC power using twisted pairs to each load. (Twisted trios or

Figure 8-10. Keeping Ground Currents Out of Your Enclosure

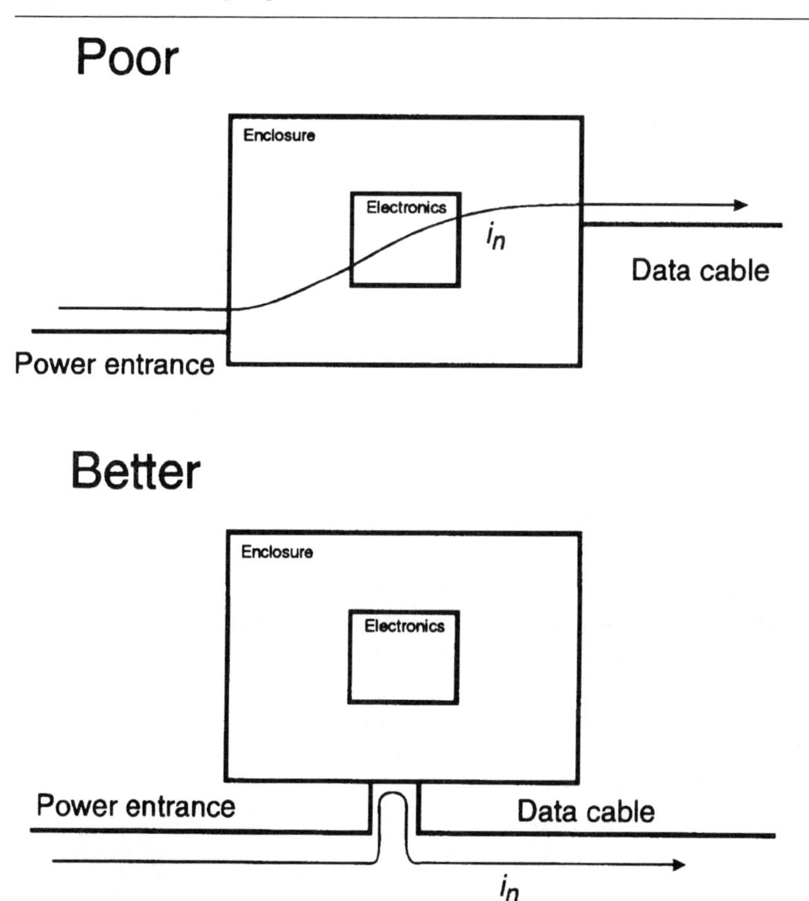

twisted quads are acceptable when multiple voltages are distributed to a module, although a twisted pair [voltage and return] for each voltage is still best.) Do not daisy chain your power between noisy loads and sensitive loads. Daisy chaining is a poor idea, anyway, and you should avoid the practice. Route these power feeds along your ground plane to minimize loop area.

Signal Cables

Route signal cables along ground plane to minimize loop areas. Avoid hanging in midair. Segregate signal cables into sensitive and noisy groups and keep the groups separate. Minimize long parallel cable runs, but if necessary, use spacers or shielded cables. Use twisted pairs to switches and indicators. Never rely on chassis for signal return.

ESD Considerations in Enclosures

Controls and indicator panels are most vulnerable. Ground all exposed metal to the enclosure with wide straps or direct connection. Panels that are exposed to touch must be very thoroughly grounded at perimeter. Shield LCD panels–unshielded panels are vulnerable to scrambling of screen contents, as a minimum. Filter or shield all lines entering the enclosure, including controls and indicators.

If you are connecting modules together in a system, we recommend mounting them on a "ground plane," and then routing interconnecting wiring close to this plane to minimize unwanted "loop areas." This will control high frequency EMI, such as digital noise or ESD. For low frequency modules we recommend a separate "single-point" ground for the "analog ground."

The ground plane provides the stable reference for all modules. This can be a base plate or a structural member, but it must meet the requirements for a low impedance high frequency ground. The arguments apply to any system, large or small.

A typical system will include a number of modules in the enclosure; perhaps a CRT, a printer, and other modules, interconnecting cables, and external cables. What rules apply to ensure a good ground for the system?

- Remember that the ground path must have a length-to-width ratio of less than five (better yet, less than three).
- Ground performance is seriously degraded if ground material is not continuous, so seams must be avoided. If it is impossible to make a ground plane solid as described, then

ground members must be very well conductively mated every few inches. Access and cable routing holes are permitted, as long as they are small relative to the size of the plane. The ground may bend and branch as needed.

- All modules must be connected to ground plane, either directly or via a short ground strap. If the module cannot be brought down to ground plane, then a ground stub should be brought up to the module (remember 5:1). The circuit boards in the module need NOT be grounded to the chassis ground.

SUMMARY

The following rules apply to subsystem interconnect:

Cabling

- Provide ample signal return paths and spread them evenly throughout the cable.
- Use filters to reduce bandwidth to only that necessary to provide adequate signal integrity.
- For high speed signals use one ground for each signal line, preferably a twisted pair.
- Differential signals are better yet.

Cable Routing

- Route cables close to continuous ground plane.
- Avoid routing cables in parallel for long runs (use rise time rule).
- Segregate cables in categories, keeping sensitive signals separate from noisy signals.
- If sufficient separation is not possible, use cable shielding.

Enclosure Design

- Ensure a continuous ground exists, suitable for routing signal and power.
- Provide a "window" for power and data entry points.
- Run twisted pair to controls and indicators.

9

SHIELDING AND SHIELDING MATERIALS

Effective shielding demands that appropriate materials be selected for the job. A common question on shielding is, "What kind of material do I need?" Actually, the shielding material is usually the *least* of concerns in most EMI applications. In most cases a good conductor is all that is needed–you can then concentrate on the most pressing issue: maintaining the integrity of the shielding material.

Let us start by discussing shielding characteristics, related to the nature of the interfering source, so that we can select the material. Then we will wrap up by discussing the main reason most shields fail: openings (including seams) and penetrations (signal and power). Good shielding decisions are not difficult to make, once the basics are understood.

THE PHYSICS OF SHIELDING

Shielding effectiveness involves two mechanisms: reflection and absorption (see Figure 9-1). In reflection an incident wave bounces off the surface; in absorption the wave penetrates the shield boundary and is absorbed as it traverses the media. Either or both mechanisms may be significant in a shielding application, but usually one will dominate.

In reflection an incident wave bounces off the surface–just like a mirror. Reflection, like a mirror, is a surface condition: What lies just below the surface is not of importance. In absorption, as with light passing through colored glass, the thicker the glass, the more the absorption. Absorption properties are a function of the media, as well as of the frequency of the wave. What makes a good reflector or absorber? The answer lies with the nature of the media and of the incident wave. Let us start with reflection.

Figure 9-1. Two Shielding Factors: Reflection and Absorption

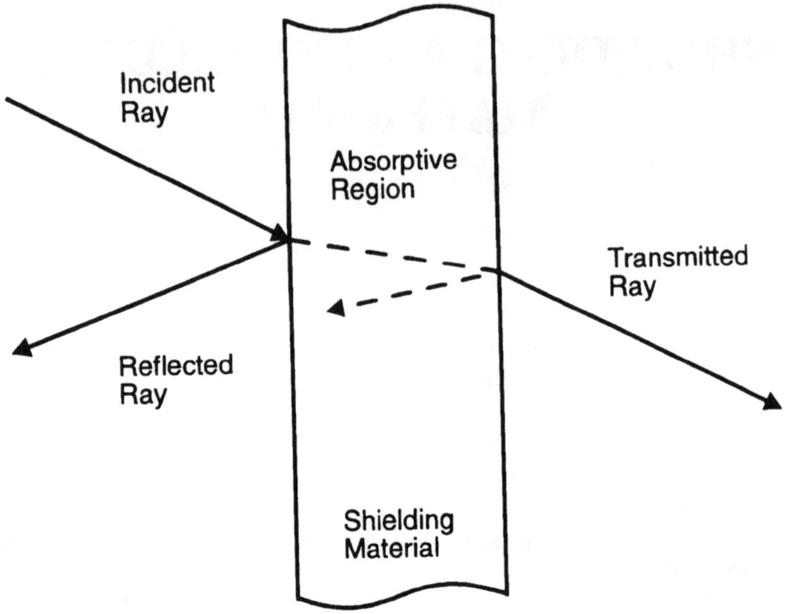

Incident Ray
Absorptive Region
Transmitted Ray
Reflected Ray
Shielding Material

Reflection depends on an impedance mismatch between the incident wave and of the reflecting surface:

$$SE = 20 \log |Z_w/4Z_b|,$$

where SE is shielding effectiveness, Z_w is the impedance of the incident wave, and Z_b is the impedance of the barrier in ohms/square. A good shield (high conductivity) would have a low barrier impedance.

The barrier impedance is in ohms/square, and must take into account the appropriate parameters at the frequency of interest. In particular, at high frequencies, conductivity occurs only near the surface, due to skin effect. Thus, above a few skin depths, the thickness of the material is of no consequence in reflective shielding.

As discussed in chapter 3 the wave impedance is the quotient of the electric field to the magnetic field immediately at the surface of the media (electric field is expressed in volts/meter and the magnetic field is expressed in amps/meter, so the result is in ohms), and varies with the distance from the source. At any significant distance from the energy source, the wave is a "plane wave" with impedance of 377 ohms. As the barrier impedance of even a moderately conductive surface is in fractions of an ohm, good shielding performance is readily achievable with most conductive surfaces.

> For example, consider a surface coating that has an impedance of 0.1 ohm/square, a conductivity that is easily achievable with most coatings and even many conductive paints. The shielding effectiveness is then
>
> $$SE = 20 \log |Z_w/4Z_b| = 20 \log (377/4 \times 0.1) = 59 \text{ dB}$$
>
> This is more than adequate for most applications.

As will be shown later in this chapter, regardless of how good the shielding effectiveness of the media is, diligent steps must be taken to control the openings and penetrations if an overall shielding effectiveness of 60 dB is to be achieved.

Conductivity cannot be taken as a constant, regardless of what is published. It is a function of the skin depth, which is given by:

$$\delta = \sqrt{(2/2\pi f \mu \sigma)}$$

where f is frequency, μ is permeability, and σ is conductivity. Most of the current passes within one skin depth of the surface, and very little current goes deeper than three skin depths. The skin depth of some common materials is given in Appendix A. As can be seen, above 1 MHz, the thickness of the material plays little part in reflection.

But close in to the source, the impedance of the wave diverges from that of a plane wave, and approaches the impedance of the circuit itself. A short dipole is a high impedance circuit, so the wave impedance near to the source is high. As distance from the dipole increases, the wave impedance decreases, ultimately falling to 377 ohms, when the short range field components in the solutions to the field equation become negligible. A small loop is a low impedance circuit, so the wave impedance close to the source is low. As distance from the loop increases, the wave impedance increases, also ultimately converging at 377 ohms. The boundary between near field and far field is defined to be at $1/2\pi$ wavelength. Closer in, you are in the "near field."

You can estimate the impedance of the wave, given the frequency of the source, the distance from the source, and the impedance of the source. We can even generalize and simply specify the circuit as high impedance or low impedance, without really worrying much about the actual numbers.

So how does this relate to shielding effectiveness? Well, earlier we mentioned that a plane wave with an impedance of 377 ohms would be well shielded by a moderately good conductor. We now need to step in and look at the impedance close in to the source. Where the impedance is high, such as from a short dipole source, the shielding effectiveness is excellent, even better than from a plane wave. Where the impedance is low, such as in a small magnetic field loop antenna, reflection is very poor. Thus, we should not attempt to shield a low frequency magnetic field with just a reflective shielding material.

Absorption, on the other hand, does not depend on the nature of the source field. It does, however, depend on frequency, as well as conductivity, permeability, and thickness. Absorption can be described by:

$$A(\mathrm{dB}) = 85 \times t \times \sqrt{\mathrm{f}\sigma\mu}$$

where f is frequency, t is thickness in mm, σ is relative conductivity, and μ is relative permeability. Note that conductivity and permeability will be frequency dependent.

As can be seen, permeability and thickness are two major factors in absorptive shielding (conductivity also plays a significant part, but conductivity does not vary as widely as permeability). We can conclude that absorptive shielding is best achieved by a thick layer of permeable material. High permeability is found in iron or other more exotic materials. The conductivity and permeability of some common materials is given in Appendix A.

So we have basically two choices when it comes to shielding: Reflective shielding (which will work very well for low frequency electric fields and for high frequency electromagnetic fields) and absorptive shielding (needed for low frequency magnetic fields). These are summarized in Table 9-1.

ESTIMATING THE IMPEDANCE OF A SOURCE

The impedance of a wave is dependent on its source, but you can generally say that the source is either high or low impedance. A crude estimate of wave impedance can be made by assuming either zero or infinite source impedance.

Table 9-1. Summary of Shielding Characteristics

Source	Wave Impedance	Reflection
Plane wave	377 Ω	High
Electric field	High	High
Magnetic field	Low	Low

For high impedance waves (dipoles, cable shields) the wave impedance is approximately (for $r < \lambda/2\pi$):

$$Z_w = 18{,}000/fr$$

where r is in meters and f is in MHz.

For low impedance waves (windings, loops) the wave impedance is approximately (for $r < \lambda/2\pi$):

$$Z_w = 7.9fr$$

where r is in meters and f is in MHz.

For example, look at a 60 Hz loop (such as a transformer) at a distance of 10 m:

$$Z_w = 7.9 \times 10 \times 60 \times 10^{-6} = 0.005 \text{ ohms}$$

This is clearly a low impedance wave.

The effectiveness of a shield for both reflection and absorption is shown in Figures 9-2 and 9-3, which is for copper and iron. As would be expected, the reflection attenuation is less for iron than for copper, due to lower conductivity. Also, note the very poor attenuation of magnetic fields. On the other hand, iron clearly provides better absorption attenuation, due to the permeability effects. As frequency increases, permeability decreases, so the performance of iron at high frequencies falls off. Permeability of metals falls off rapidly above 1 MHz, but is still providing some benefit up to about 1 GHz. Also note that, where reflection effectiveness goes down at increasing frequency, due to decreasing skin depth, absorption is increasing, and becomes significant at high frequencies where even a thin layer gives many skin depths of absorption.

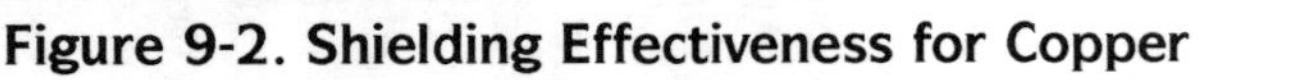

Figure 9-2. Shielding Effectiveness for Copper

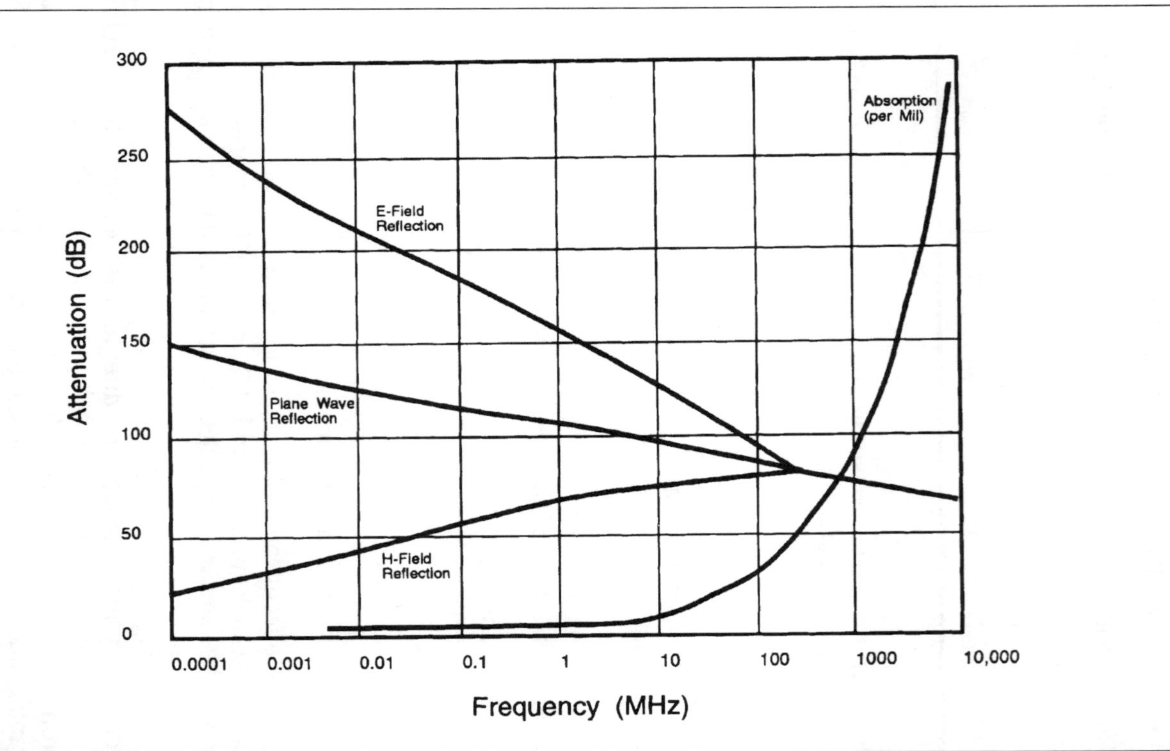

Figure 9-3. Shielding Effectiveness for Iron

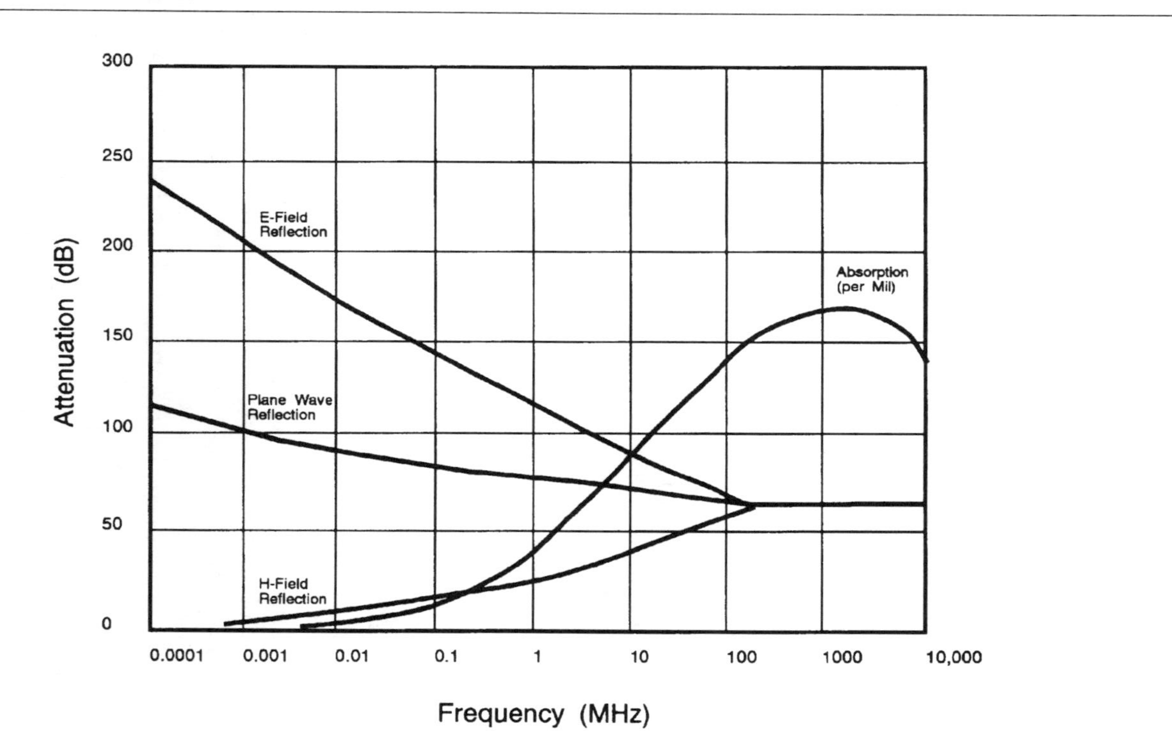

SHIELDING REQUIREMENTS

Once you have identified the nature of the source, you must decide how much shielding effectiveness is needed. You need to look at the threat to make this determination. First, is our equipment the source or the recipient of the interference? If you have a particularly noisy source, you are looking at keeping interference within your equipment. If you have a particularly sensitive instrument, you are looking at keeping interference away from your equipment. In both cases you need to assess the external environment as well. Your neighboring equipment may be either very noisy or very sensitive. In any event your shielding requirements are simply the ratio of the incident field strength to the allowed field strength, or, in log terms, $SE = 20 \log (E_1/E_2)$.

The effectiveness of high frequency shielding materials can easily be 200 dB or more; however, due to shield penetrations, shielding effectiveness of 60 dB or more requires careful design. In fact, due to these same conditions, shielding effectiveness of more than 90 dB in a single enclosure is difficult. If you need more than this, look to multiple shields. Absorptive shielding depends on thickness, and even 60 dB of shielding for low frequencies is an ambitious goal.

SHIELDING MATERIALS

Reflective Shielding

Thickness is not a driving factor for reflective shielding; surface conductivity is the issue. At high frequencies skin depth is so thin that only the surface takes part in reflective shielding. Adequate shielding can be obtained by using more conductive metals and many conductive coatings. As effectiveness varies directly with surface conductivity, we need to look at this parameter. For a given thickness copper is about the best of the readily available enclosure materials. The conductivity is usually stated relative to copper (or conductivity = 1). Conductivity of various conductive materials is given in Appendix B. Note that for most cases the shielding effectiveness of copper is far more than needed, so lesser conductors, such as mild steel or aluminum, are entirely adequate for shielding. Due to skin depth considerations, high frequency current travels within a few skin depths of the surface. As can be seen in Table 9-1 (on page 173), thickness of the media is not significant even as low as 100 kHz.

As surface conductivity is the driving factor in reflective shielding, a conductive coating over plastic can be considered. Two common processes, electroless plating and vacuum plating, provide for shielding effectiveness of 60 to 80 dB, which is enough for most applications. Electroless plating usually uses nickel and is commonly used in electronics today. Vacuum plating usually uses aluminum and is often preferred for short runs. Both are complex processes and are usually jobbed out.

Conductive paints are usually about 80 percent metal (silver, copper, or nickel) and 20 percent binder, giving a shielding effectiveness of 40 to 60 dB, which is still adequate for many applications. Conductive paints are available in spray cans and can be easily applied in the lab or field. Conductive fibers thermoformed over plastic provide about the same performance as conductive paints, but are much more durable (Figure 9-4).

Figure 9-4. Shielding Thermoformed Over Plastic (Photo courtesy of 3M)

Conductive plastics are filled with metal fibers or flakes. Shielding effectiveness is typically about 40 dB, which is still enough for many applications. However, there is considerable difficulty in mating adjacent conductive surfaces, due to poor surface conductivity; this makes it very difficult to actually achieve the desired shielding effectiveness, so we are decidedly lukewarm on this process.

There are also other marginally conductive coatings, such as carbon or graphite. These are acceptable for draining ESD currents from the enclosure, but are inadequate for most EMI applications.

Absorptive Shielding

Permeable materials are most appropriate for absorptive shielding. Mild steel, such as found in sheet metal, has good permeability. Note that while the conductivity of steel is only about 1/6 that of copper, the permeability more than makes up for the deficiency. If you go with some of the exotic magnetic materials (like mumetal), you can obtain much higher permeabilities, yet.

When selecting the magnetic material, many conditions will reduce the effective permeability of the shield. The shield needs to be a closed path, otherwise shield effectiveness is significantly reduced. Welded seams are preferable, but heavily lapped seams are often good enough. The idea is to reduce the reluctance of the path to a minimum.

All permeable materials will saturate if subjected to high enough magnetic field. If saturation is a concern, a high permeability material can be buffered with a lower permeability material facing the field source. This could be accomplished with laminations.

Mild steel can be worked easily, but exotic permeable metals cannot. They are annealed at high temperatures in an inert atmosphere. Kinking or drilling holes will significantly reduce permeability.

If you are not experienced in handling high permeability materials, you may want to have your fabrication done by a specialty house. By the way, high permeability materials are quite expensive, so do not use these materials unless it is absolutely necessary.

Plating

In most applications the shield will be treated for corrosion protection. Generally, coatings will not adversely affect shielding properties. In fact, high conductivity coatings will enhance poor conductors; shielding effectiveness can then be determined primarily by the coating. In any event it is imperative that the mating surfaces be conductive. Specifically, anodizing and paint are not conductors. Table 9-2 gives some suitable coatings for aluminum and steel. Not all of these coatings are durable, so be sure to consider how often your surfaces will be mated. If the mating surfaces are only separated during repairs, the surface need not be very durable. If the mating surfaces are separated often, as in the case of a cabinet door, the surfaces need to be considerably more durable.

Materials Compatibility

As long as good conduction is maintained at the mating surfaces, the shield will perform well, and this is generally true even for dissimilar metals. Unfortunately, surface conditions deteriorate over a period of time. Two phenomena contribute to degradation of the contact: electrolytic corrosion and galvanic action. This was discussed in chapter 6 on Grounding and applies equally well to shielding.

Galvanic action occurs as a result of the mating of two dissimilar metals, which gives rise to currents. Corrosion occurs in the presence of an electrolyte (usually moisture), primarily at one of the two metals. The approach is to use metals close together on the electrochemical series (Appendix A). Note that aluminum is not closely matched to iron, and is not a good match (camper enthusiasts with aluminum toppers have found this out). The solution is to use metals close on the electrochemical series, or to eliminate moisture at the contact area. By the way, this effect is a surface condition, so plating with more compatible metals is one way to obtain more compatible surfaces.

Electrolytic corrosion occurs when current flows through a contact in the presence of an electrolyte (again, moisture), and this

Table 9-2. Suitable Coatings for Aluminum and Steel

Aluminum	Steel
Irridite	Zinc Chromate
Clear Chromate	Zinc
Tin	Cadmium
Alodine	Nickel

Conductive paints will also work on either steel or aluminum.

occurs even if the mating metals are the same type. The solution is to block moisture from mating surfaces.

WHERE SHIELDING FAILS

Shielding failures are usually due to breaches in the shield and not due to the material itself. These include shield openings and shield penetrations (Figure 9.5). Shield openings are necessary in any practical enclosure: Seams and ventilation holes represent the biggest problems. In chapter 3 we discussed the slot antenna, where the slot effectively interrupts the laminar current flow on one or both surfaces. The key issue in the opening is the longest dimension, whether it is a diameter, a diagonal, or simply the length of a long slit. It is surprising to many that a thin slit in a seam is as bad a leak as a circle of the same diameter, but it is a fact. Note that the issue is not opacity, but the current path, so lapping seams does not significantly alter the arguments. It must be presumed that the space between the fasteners is essentially open and nonconductive.

Shielding effectiveness of an opening is given by:

Figure 9-5. Shielding Failures

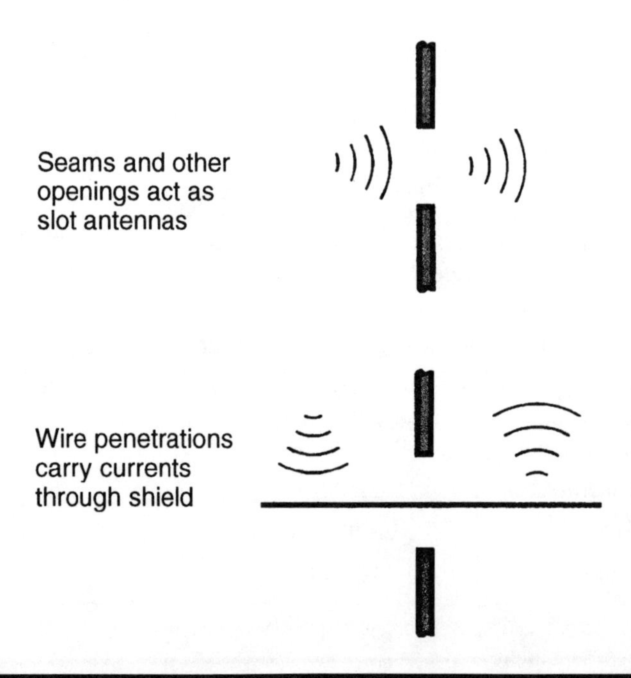

$$SE = 20 \log (\lambda/2L)$$

where λ is wavelength and L is longest dimension in the opening. We consider a 1/20 wavelength as about the minimum useful, as it gives a 20 dB shielding effectiveness. Anything less is hardly worth considering.

> Let's apply this to a real world problem. Suppose we have an immunity requirement up to a frequency of 1 GHz (wavelength 30 cm). A 1/20 wavelength slot is 1.5 cm or a little over 1/2 inch. Thus, if we want a 20 dB shield, we need to ensure that there are no openings no larger than 1/2 inch long. This is quite easy to achieve with ventilation holes, as a series of perforations will serve, but it does pose a problem with the seams. Do you really want to place your fasteners 1/2 inch apart to get a mediocre shield? Obviously, this is very impractical. Accordingly, we resort to EMI gaskets, which bridge the gap between the mating surfaces.

GASKETING

As mentioned, when two mating surfaces are pinned together, the two surfaces scallop (Figure 9-6), leaving a gap. EMI gasketing is

Figure 9-6. Mating Surfaces Scallop

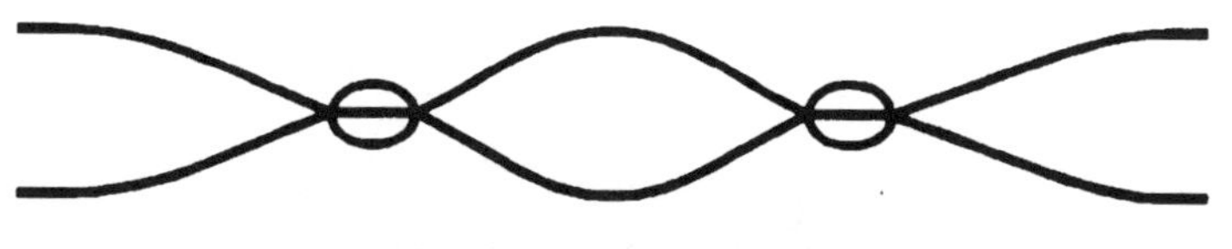

Surfaces scallop when fastened, leaving open slots between fasteners

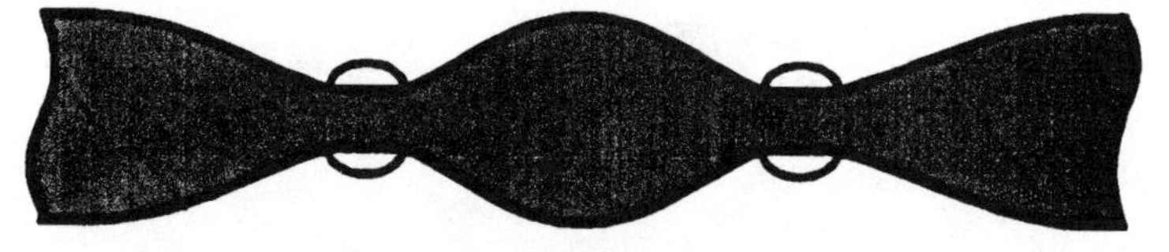

Slots can be closed using resilient metal gasketing

used to close the openings between two mating surfaces. This places two constraints on the surfaces. First, it must be conductive so that the conductive gasket can conductively close the seams (no paint or anodizing at the mating surfaces). The second is that the surface stiffness be such that the gasket can effectively close the seams.

There are a number of effective gasket materials, and the selection is primarily for mechanical reasons. Generally, any available EMI gasket has sufficient conductivity, provided that the seam has been closed. The common gasket materials are characterized in Table 9-3.

Finger stock is generally the most resilient of the gasket types, being composed of a metallic strip with "fingers" cut into one side (Figure 9-7). Finger stock is extremely forgiving of compression, and

Table 9-3. Common Gasket Materials

Finger Stock
20–80% compression range
Very low closure pressure
May close in shear
No compression set
Vulnerable to snags
Use for large doors and openings
Use in small strips to close slot
Wire Mesh
With or without elastomer core
60–80% compression range
Moderate closure pressure
Will set if overcompressed
Use for medium size doors and panels
Conductive Elastomer
85–95% compression range
High closure force
High compression set
Use for connectors and small panels

Figure 9-7. Finger Stock Gasketing (Photo courtesy of Instrument Specialties)

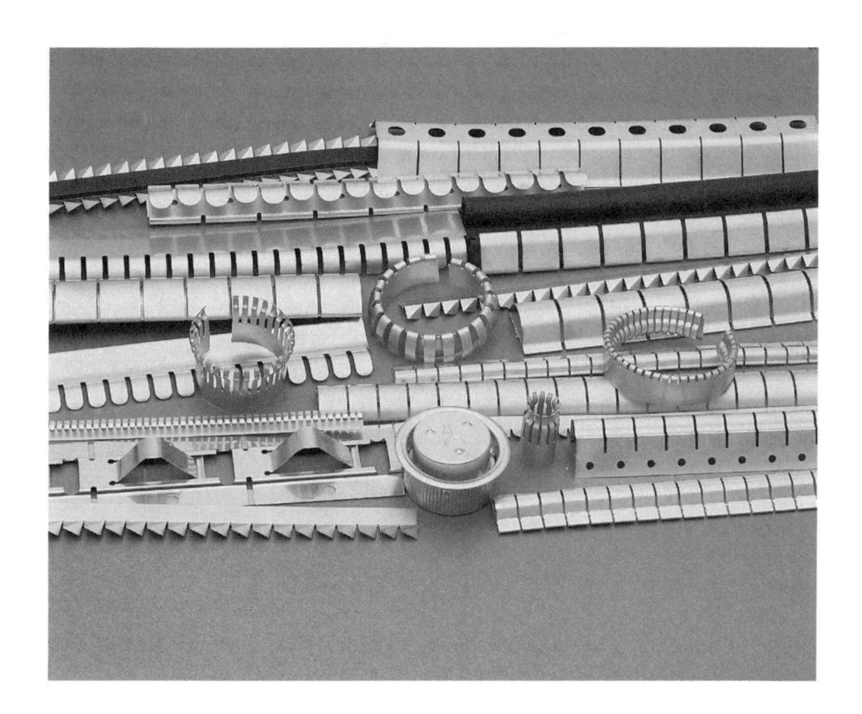

is the only gasket than can be closed in shear. These fingers are very vulnerable to snagging by humans. A similar technology uses a spiral wrap. Although less versatile in styles, it does eliminate the problem with snags.

Knitted meshes are composed of woven metal wires (figure 9-8), and may be center filled with an elastic foam to provide for greater flexibility. They are less flexible than finger stock (although some are pretty flexible) and less forgiving of compression. Closure pressure is higher.

Conductive elastomers have metal filaments, providing conductivity between the two mating surfaces, embedded in an elastomer (Figure 9-8). They are quite stiff and are usually used only for small

Figure 9-8. Knitted Mesh Gasketing and Elastomer Gasketing (Photo courtesy of Chomerics Division of Parker Hannifin)

openings, such as for connectors, although they are also used where the surfaces can be screwed together at frequent intervals.

Any of these materials will provide effective shielding if properly installed: the key is essentially continuous closure. Normally, the gasket material is selected for compatibility with mating surfaces. Do not let the gasket salesperson dazzle you with a list of exotic materials to select from. With few exceptions (notably magnetic shielding), all will meet your needs as long as the gasket is reasonably conductive over the long term. If you are looking for magnetic shielding, then it is appropriate to look at a gasket with a permeable material.

The question is, inevitably, "Do I really need gasketing?" That depends on your shielding needs. In many cases your needs are very modest, and a metal enclosure with an occasional fastener may be sufficient. If you get by without gasketing, then you have a case where you did not really need much shielding. But if you have significant shielding needs, you will not achieve success without nearly full gasketing.

PERFORATIONS AND SCREENS

For the most part perforations and screens both perform well in shielding applications. If the opening is small relative to a 1/2 wavelength, the shield loses some effectiveness due to reduced conductivity. But in most cases, the shielding effectiveness is so high that you can afford to give up a lot. If the shielding effectiveness of your material is 200 dB, what difference does it make if you sacrifice 60 dB or so? Only when the frequency gets high enough that the opening becomes significant (relative to a wavelength) does the shield start to degrade significantly. Perforated openings are effective in closing ventilation holes, as shown in Figure 9-9. If additional shielding effectiveness is really needed, a honeycomb screen can be used, keeping the shielding effectiveness very high. This is a drastic

Figure 9-9. Shielding Effectiveness of Perforations

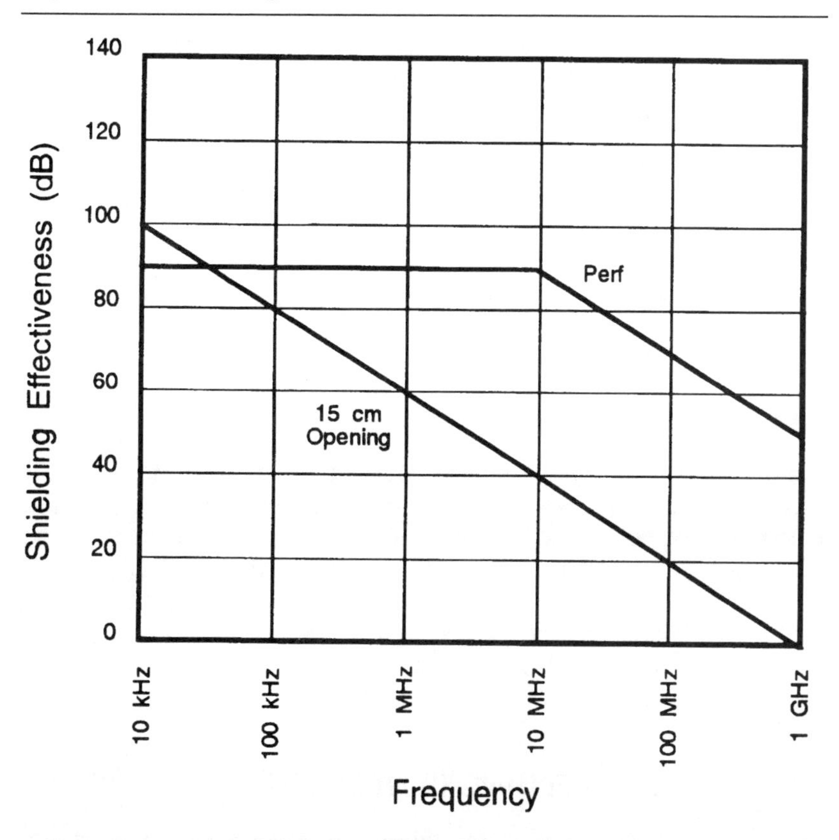

measure, as even a small honeycomb screen will cost $100 or more, and it still must be very carefully mounted.

Screens would act essentially the same as perforated metal, except that the intersections are not solidly conductive. Screens are still pretty good up to about 300 MHz, however, when the degradation starts to become significant. To be effective, the screen must be bonded around the perimeter rather than just at the corners.

CABLE PENETRATIONS

Any conductor that penetrates the shield will pick up energy from one side, carry it to the other side, and either reradiate it on the other side or carry it directly to the recipient circuit. It actually takes but little, just a short length of unconnected wire will completely destroy the shielding effectiveness of the best shielded enclosure.

How do we cure this? All high frequency energy must be shunted off to the enclosure boundary. There are precisely two ways of handling shield penetrations, one for shielded cable and one for unshielded cable. For shielded cables the shield must be terminated to the boundary. For unshielded cables the line must be capacitively terminated to the boundary (Figure 9-10). ***Any energy that is not shunted off to the case will penetrate and destroy the shield.*** Obviously, this poses some severe restrictions on design, as leakage current limitations will often prevent using adequate amounts of capacitance.

GROUNDING THE SHIELD

We often run into the case where the shield has been run to earth ground. (Sometimes we find the shield has been run to a separate earth ground, sometimes running hundreds of feet. Fortunately, we have not yet encountered such a case in the medical world). Grounding the shield does not improve shielding effectiveness in the least: Terminate the cables at the enclosure boundary, and ground the enclosure as appropriate for safety requirements. Shield terminations are covered in chapter 10.

HOW MUCH SHIELDING IS NEEDED?

There is no certain way to determine analytically how much shielding is need. One way to tell is to test the equipment with no

Figure 9-10. Terminating Penetrations

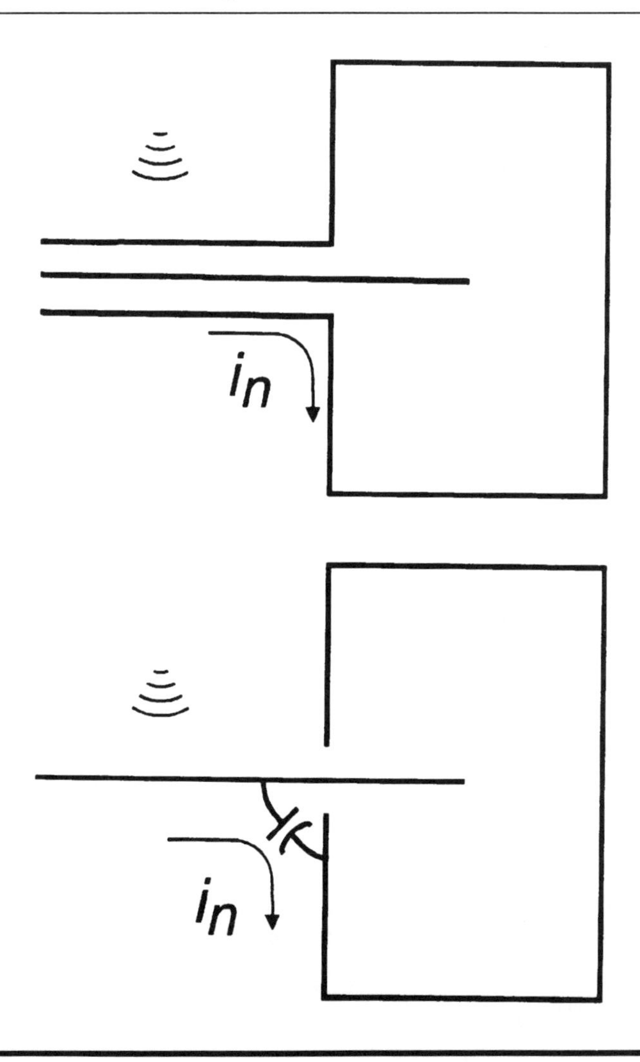

shielding. Once the deficiency is determined, one can design an enclosure to meet or exceed the needs. This can backfire with emission tests, however. An unshielded box will radiate in all directions, having no particular concentration in any direction. Mounting a shielded enclosure, however, will columnate the energy into specific directions, sometimes actually raising the maximum emission levels.

The real problem with the prediction of shielding needs is that there is such a wide variety of equipment, of design practices, and environments. We can say, however, that the better the internal electronics are designed, the lower the demand on the enclosure. We know that good internal design practices will easily provide a 20 dB (10 times) improvement over casually designed equipment, reducing shielding needs by the same factor.

For reference we consider that a commercial enclosure will typically need about 30 to 40 dB of shielding, a military enclosure 60 dB or more, and some high interference environments more yet. Generally, we consider a 20 dB shield as being a minimum goal.

In many cases medical electronics are mission critical, and susceptibility requirements may be imposed. If a 10 volt/meter susceptibility requirement is levied, then a 40 dB shielding effectiveness will reduce the internal field to 0.1 volt/meter, which is probably good enough for all but sensitive analog circuits. These, of course, can be dealt with to a significant extent by onboard remedies as described in previous chapters.

PARTIAL SHIELDS

The goal of shielding is to block adverse energy from passing in or out of equipment boundaries. Never forget, however, that there are other ways of minimizing energy transfer. Good circuit board design and good internal cable layout will minimize the need for shielding. If a full shield is not possible or practical, then one can make maximum use of existing shielding or ground members.

The most critical location on an enclosure is at cable entry, and where internal cables lie close to the seams (see chapter 8). Cable and power entry should be on a single plate, or "window," with no seams in the immediate proximity of the cables. Leave as much clearance as possible, but at least several inches. This means, specifically, that there are NO seams in the window.

LOCAL SHIELDS

Understand that the above shielding requirements may be difficult or impossible to achieve. Where it is impossible to shield or to capacitively filter a line, an alternate may be to use a local shield. Radio engineers have long used local shields for specific circuits on a circuit board, and this practice is transferable to other pursuits.

Particularly where sensitive analog circuits need to be protected, you can filter the circuit to an isolated "doghouse," a local shield on the circuit board, and connected to circuit ground. The amplified output of the analog circuit will be at significantly higher levels, and much more capable of withstanding interference than the input circuit. Obviously, this does not retain the integrity of the overall shield, as the cable penetrations still exist, but you will reduce the susceptibility of the most sensitive circuits. Actually, the doghouse can be extended to encompass all of the electronics, either in one large house or in "condominiums," where there are many separate units. Thus, when looking at shielding, look carefully at how selected shields may be used to full advantage, while maintaining the necessary electrical isolation.

SUMMARY

We can summarize the shielding requirements with a few simple rules:

- Thin metallic material of almost any type (aluminum or iron sheet metal, conductive platings, conductive spray paint) will serve for high frequency shielding or low frequency electric field shielding.

- Low frequency magnetic fields such as needed for power supplies need thick permeable materials, either mild steel or one of the high permeable materials.

- The principal reasons for shielding failures are shield penetrations and openings.

- Openings must be minimized in order to achieve good shielding effectiveness (use the 1/20 of a wavelength criteria). This includes ventilation holes, seams, and so on.

- Gasketing materials can be used to close seams, continuous where high shielding effectiveness is needed or in separated strips if mild shielding effectiveness is acceptable. Gasketing should be selected for mechanical reasons–whichever one is the one most feasible to implement mechanically.

- All energy on wires and cables must by terminated at the shield, either by terminating the cable shield to the enclosure shield or by capacitively terminating the wire to the shield enclosure.

10

CABLES AND CONNECTORS

Cables and connectors are very important players in the EMI game. They are major sources of EMI leaks into and out of electronic systems. While cables are supposed to carry intended energy or information, they also carry unintended energy that can cause EMI problems. At high frequencies they act as unintended antennas for radiated EMI; at low frequencies they act as unintended conduits for conducted EMI. Remember, that a "low frequency" patient cable or power lead can pick up or radiate energy well into the gigahertz range.

Cables must be designed as a system. They are a sophisticated collection of conductors, shields, and filters that must be properly designed, both mechanically and electrically. Successful EMI designs pay attention to the cables, while unsuccessful designs do not. Most radiated EMI problems, and many conducted EMI problems, have cables as a major contributing factor.

In this chapter we will concentrate on cable shielding and terminations external to the enclosure. Also refer to chapter 8, where internal cable design is discussed, much of which is also relevent to external cabling. We will look at a range of cable-related EMI problems, and provide some EMI design guidelines on cable construction, shielding, and cable connectors, an often overlooked but very critical part of any cable design.

CABLES AS ANTENNAS

Any conductor will act as an efficient antenna when its physical dimensions exceed a fraction of a wavelength. Cables, by their very nature, are among the longest conductors in a system, so they are usually the most significant "antennas" as well. If you understand the antenna effects of cables, you will understand how to prevent and solve many of your high frequency EMI problems.

As a rule of thumb, any wire over 1/20 of a wavelength is considered an antenna for EMI purposes. Most communications antennas are 1/4 of a wavelength or more in length, but even "short" antennas can be significant radiators. Table 10-1 shows some typical frequencies and wavelengths, which are related by the formula $\lambda = 300/f$, where λ is the wavelength in meters, and f is the frequency in MHz. Also shown in this table are the "equivalent" frequencies of digital signals, related by the formula $t_r = (1000 \times f)/\pi$, where t_r is the rise time in nanoseconds and f is the frequency in MHz.

> For example, a cable that is 1.5 meters long is a good antenna for any frequencies above about 10 MHz. Even a 50 cm cable is a good antenna above 30 MHz. Since most radiated problems (both emissions and immunity) occur in the 30 MHz–1000 MHz range, it is clear that cables are major contributors. Incidentally, at frequencies above about 300 MHz, note that the critical dimension drops to 5 cm. Thus, at frequencies above 300 MHz, even the board traces themselves become significant radiators.

As a rule of thumb, we usually assume the following:

- $f < 30$ MHz Most likely EMI path is conduction (Cable effects minimal)
- 30 MHz $< f <$ 300 MHz Most likely EMI path is cable radiation
- $f > 300$ MHz Several likely paths–cable radiation, board radiation, enclosure radiation

Table 10-1. Typical frequencies and Wavelengths

Frequency	t_r	Wavelength	1/20 Wavelength
300 kHz	0.96 μs	1000 m	150 m
1 MHz	287 ns	300 m	15 m
3 MHz	96 ns	100 m	5 m
10 MHz	29 ns	30 m	1.5 m
30 MHz	10 ns	10 m	50 cm
100 MHz	3 ns	3 m	15 cm
300 MHz	1 ns	1 m	5 cm
1000 MHz	0.3 ns	30 cm	1.5 cm

COMMON MODE VS. DIFFERENTIAL MODE

As has been discussed in chapter 8, CM and DM interference both exist side by side, and each must be addressed, as the remedies are essentially separate. Generally, CM interference dominates at any distance from the source. For shielded cabling interference is essentially all CM, and, even for most unshielded cables, CM is still dominant.

A rather unique situation exists with patient cables, however, where both CM and DM are significant. This will be discussed later in the chapter.

BANDWIDTH

A key issue when designing cables is determining the necessary frequency range. This includes both the intended frequency range and the threat frequency range. For example, an AC power cable only needs a 60 Hz bandwidth, but it may be exposed to radio frequency (RF) threats well into the gigahertz range. On the other hand, if nearby transmitters are not a threat, you may still be affected by nearby 60 Hz electric or magnetic fields.

The intended frequency range is a very important parameter for shielded cable design. The frequency affects not only the shield material, but also the grounding of the shield and the connector types and construction. Many EMI problems are due to using a low frequency shield for a high frequency application, and vice versa. In some cases separate low and high frequency shields may be necessary.

Knowing the bandwidth lets you make trade-offs on cable shielding versus filtering. If the intended frequency range is low (such as power lines or low speed signal interfaces) and the threat frequency range is high (RF emissions, RF immunity, ESD), then high frequency filtering can be used to block unintended energy coupled on the cable at the point of entry. If, on the other hand, both the intended frequency range and threat range are high, then shielding must be used, since filtering would block or degrade the intended signals. In this case we must prevent the energy from ever reaching the vulnerable lines in the cable.

A similar situation exists with low frequency analog interfaces (such as ECG inputs) and low frequency threats (such as 60 Hz power line fields). Once again, since the threat frequency is similar to the operational frequencies, filtering is not practical, so shielding must be used. In some cases, however, 60 Hz "notch filters" can be used.

Another common low frequency technique is to use balanced circuits with high CM rejection. Since cable pickup is primarily CM (patient cables excepted), this can be quite effective. This works best with low frequency threats such as 60 Hz, since as the threat frequency increases, CM rejection decreases due to parasitic capacitance "unbalancing" the circuit.

Remember, you need to determine both the operational frequency range and the EMI threat frequency range of any cable. This knowledge will help you decide whether to use filters or shielding to block cable-related EMI. As we will see later in this chapter, these frequencies will also determine how good the cable shields and connectors need to be.

CLASSIFYING CABLES

Because cables differ so widely in their intended applications, it is helpful to classify them by bandwidth. Table 10-2 is an example of a classification scheme we developed several years ago for a telecommunications client. In this case there was much confusion about the "best" way to treat cables. Unfortunately, there is no single "best" way—the preferred approaches depend on both the bandwidth and the threat frequencies.

Note that at the highest frequencies (Class I), no filters were used since any filtering would likely degrade the intended signal. In this case the entire EMI strategy was to shield the cable, and do a very good job. For medium frequencies (Class II and III), filters and shields were used together. At low frequencies (Class IV, V, and VI) only filtering was used.

Note also the comments on shield grounding. As a rule of thumb, we use single-point grounding of cable shields below 10 kHz, and multipoint grounding above 10 kHz. In special cases we will use a "hybrid ground" with capacitors on one end, to provide grounding on both ends for high frequencies, and one end for low frequencies. We will look at this in more detail later in this chapter.

CABLE SHIELDING AND CONNECTORS

Although many cables are shielded, most designers are confused about the best way provide this shielding. Many questions are raised, and many opinions abound. Is a high quality shield really necessary? Should I ground the shield at one or both? And where is the best place to ground the shield? Do I need high quality metal connectors? Are pigtails on the shields permissible?

Table 10-2. Cable Classification Scheme

Class	Use	Bandwidth	Filter	Shield/Connector	Ground
I	High speed communications	> 100 MHz	No	Very high quality	Both ends
II	Medium speed comm (LAN)	10 MHz	> 10 MHz	High quality coax	Both ends
III	RS232 (9600 baud)	10 kHz	> 100 kHz	Medium quality metal connectors	Both ends
IV	Audio	3 kHz	< 300 Hz and > 3 kHz	Shielded twisted pair	One end
V	Power	60 Hz	> 10 kHz	Twisted pair	N/A
VI	Alarm, indicators	< 5 Hz	> 10 kHz	Twisted pair	N/A

There is no single "right" answer to any of the above questions. The correct answers depend on a number of parameters: the particular application, the operational frequency range, the threat frequency range, the sensitivity of the victim circuits, and patient safety considerations. In some cases the preferred approaches even conflict, which leads to trade-offs or hybrid solutions.

Three Shielding Needs

To begin with, it helps to understand that we are actually dealing with three different cases where shielding is needed, which we can explain with three different models. The first two models (capacitive and inductive) help explain the behavior of low frequency, or "electrically short," shields. The third model (electromagnetic) helps explain the behavior of high frequency, or "electrically long," shields.

For medical devices, we are interested in all three cases. The capacitive model is preferred when dealing with high impedance analog circuits when we are shielding against power line electric fields. The inductive model is preferred when dealing with low impedance analog circuits when we are shielding against power line magnetic fields. The electromagnetic model is preferred when using shielding to contain RF emissions or to enhance RF immunity.

Incidentally, the transition between "electrically long" and "electrically short" shields occurs at about 1/20 of a wavelength. For shorter cables shorter cable shielding can be predicted using lumped component models. For longer cables longer cable shielding must account for transmission line and antenna effects. As we saw earlier in this chapter (Table 10-1 on page 192), any cable over 1.5 meters in length is "electrically long" at frequencies above about 10 MHz. Thus, most of our external cables are electrically long in the RF range. On the other hand, at 60 Hz, a cable is "electrically short" if it is less than 800 kilometers long. Clearly, most cables used in medical devices are "electrically short" at power line frequencies.

Low Frequency Shielding

Figure 10-1 shows the shielding model for low frequency electric field shielding. This is also probably the easiest model to understand, so it is a good starting point. This model works very well for high impedance circuits and low frequency threats, such as ECGs bothered by 60 Hz electric fields.

In this model the shield simply intercepts the capacitive coupling from "space" to the wires, by putting another capacitor (the

Figure 10-1. Model for Low Frequency Electric Field Shielding

Capacitive coupling to cable

V_n

V_n

Equivalent circuit

Cable shield grounded at Load

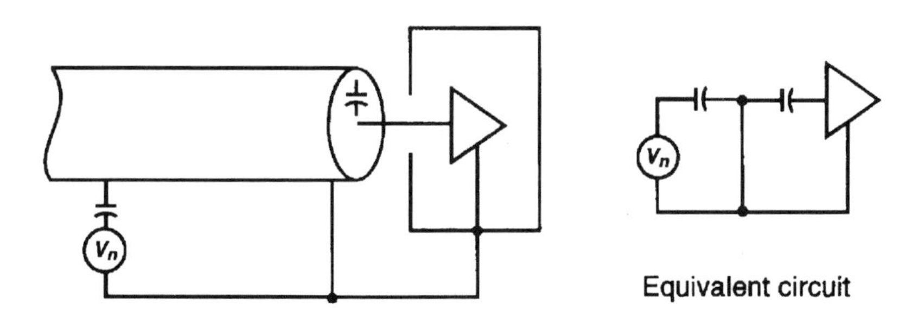

shield to wire) in series. If the shield is then grounded, any capacitive currents are intercepted, and shunted away from the circuit. The secret to success is maintaining a low impedance connection to the ground point. If the impedance is too high, the energy is not diverted, but passes through the two capacitors.

In order to make sure that we maintain a low impedance ground, we must consider its impedance. For frequencies below about 10 kHz, the resistance of a wire is very low, so a "pigtail" connection provides a satisfactory ground. At frequencies above 10 kHz, however, the wire impedance becomes inductive, and thus it increases rapidly with frequency. As a result, *for frequencies above 10 kHz, pigtail connections should not be used to terminate shields, even if they are "electrically short."* We cannot overemphasize this point–it is a leading cause for cable shielding failures.

Note that for electrically "short" cables, the shield is at the same potential all along the shield, so only one ground point is needed. In fact, the single-point ground is preferred for low frequency high impedance applications, as it reduces the potential for ground loop problems.

As the cable length increases (or the frequencies increase), the cable eventually becomes electrically long. In this case the single-point ground no longer works. Even if we have a perfect ground at one end, a high impedance point will exist on the "open" end due to transmission line effects. This allows energy to couple between the cable and shield, which will destroy the shielding effectiveness of the cable shield. Thus, ***for electrically "long" cables, the cable shield must be grounded at both ends.*** And you must use the threat frequency, not the operational frequency, to determine the electrical length.

> For example, let us assume we have a low ECG interface using a shielded cable. To protect against 60 Hz electric fields, we ground the shield at one end (at the amplifier neutral is preferred). On the other hand, to protect against RF fields, we either need to ground at both ends, or we need to incorporate high frequency filtering at the input of the circuit. The single-point shield in this case is totally ineffective against RF energy. (Incidentally, we routinely run into this problem with medical devices.)

Another case requiring grounding on both ends is low frequency magnetic field shielding. In this case the magnetic coupling is decreased by providing a nearby path for the return current to flow. This minimizes the loop size, and thus the magnetic pickup or radiation. Figure 10-2 shows the shielding model for low frequency magnetic field shielding. This model works very well for low impedance analog circuits that are vulnerable to power line magnetic field interference. In this model the shield transforms the inductive coupling by providing a shorted secondary (the shield), protecting the line within. Note that in this case, both ends of the shield must be grounded, which conflicts with the above electric field shield requirement.

Another type of magnetic shielding uses high permeability materials, such as steel or mumetal, to concentrate the magnetic flux and thus minimize magnetic coupling. In this case no ground is required–the objective here is to minimize the "magnetic path" of the magnetic flux. This is sometimes referred to as ducting, and is shown in Figure 10-3.

Figure 10-2. Model for Low Frequency Magnetic Field Shielding

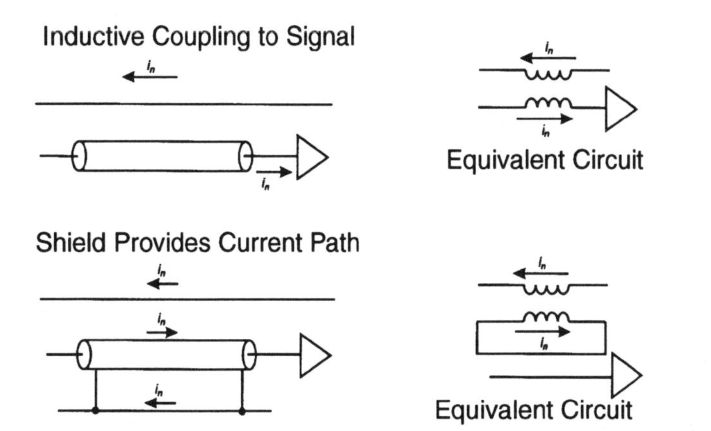

Figure 10-3. Magnetic Field Ducting

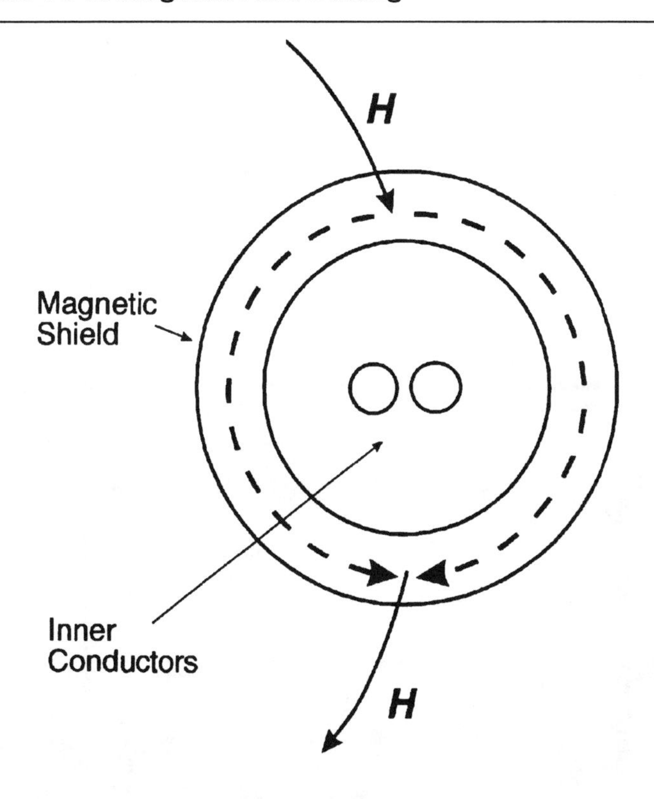

High Frequency Shielding

Figure 10-4 shows the model for high frequency shielding. This is often referred to as the barbell approach to shielding. The idea is to fully enclose the circuits on both ends of the cable, as well as the cable itself. The key issues here are not to have any breaks in the shielding, and to provide enough shielding effectiveness in the cable shield itself. The first issue deals primarily with connectors and the second with the "transfer impedance" of the shield.

Most high frequency shielding problems occur at the connectors. As mentioned earlier, pigtail connections should not be used for high frequency shielding, due to the high inductance. Figure 10-5 shows the equivalent frequency of a pigtail connection–it is almost like the matching network on an antenna! Pigtails will ruin even the best cable shield at RF and ESD frequencies.

Figure 10-4. Model for High Frequency Cable Shielding

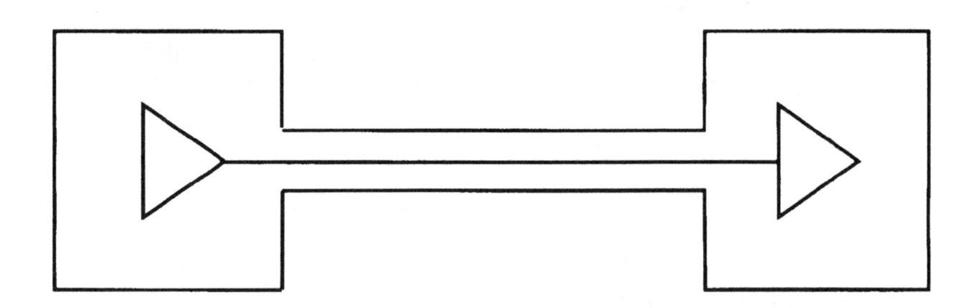

Figure 10-5. Equivalent Circuit of Pigtail Connection

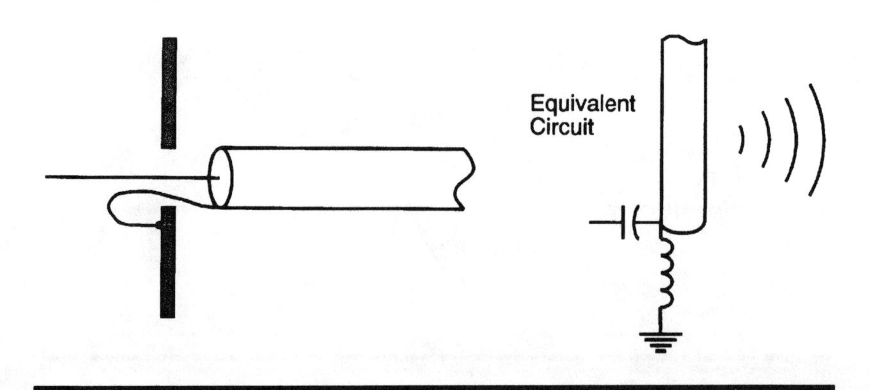

CABLE TERMINATIONS

The principle reason that cables fail is because of poor cable terminations. Where leakage current limits do not force compromises in shield terminations, the terminations are straightforward: Either the shield or the filter is terminated to the enclosure, thus shunting interference currents. Such practice is effective and widespread in industry, and is applicable in diagnostic equipment (but not for patient-connected devices.)

Connectors have many weak points, as shown in Figure 10-6. You must pay attention to all of the joints. As system speeds or frequencies increase, even a small discontinuity can cause major

Figure 10-6. Leakage Paths in Cable Connector

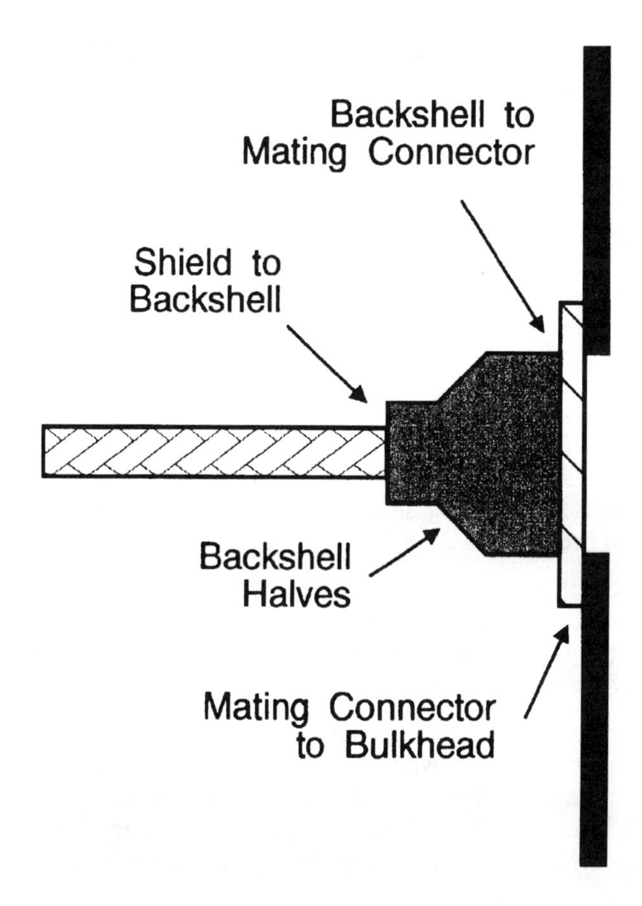

shielding leaks. It is like a garden hose—even the best hose will leak if the connection is not tight at the faucet. So it is with high frequency cables.

At the *cable-to-connector* connection you need a full circumferential bond at the shield. The better the connector, the better the method of gripping the shield, so high quality connectors are a must for high frequency cable shields. This is difficult to achieve with sheathed Mylar® film cables, where the foil cannot readily be formed to the connector. The *connector-to-connector* joint is important, and must be metal to metal, not plastic. At frequencies above about 500 MHz, dimpled connections are preferred over smooth connections. Some connectors even include fingerstock gasketing to maintain high frequency continuity at this vital connection.

Finally, do not overlook the connector to chassis connection. One common problem is chassis connectors that do not overlap with the chassis at the cutouts. This imprecise fit creates a slot between the connector and chassis that leaks and couples energy onto the shield. If you cannot use "fat" connectors here, several vendors make special gaskets to seal this connection.

Patient Cables

This is the tough one, as it forces a violation of almost all good EMI design practices:

- Cable shield termination to enclosure may be restricted because of leakage current limitations.
- Cable shield ends at the patient end, preventing grounding of both ends of the cable.
- The cable usually splits near the patient, and this portion forms a DM path, as well as a CM path outside the shield (Figure 10-7).

As such, the effectiveness of cable shielding is severely limited, and extreme measures are required elsewhere. The options are limited, and they are presented here, not necessarily as a usable remedy, but to steer you in the right direction:

- At the patient end where the cable splits, place a high frequency filter network at each signal line, terminating each line with a high frequency capacitor to the cable shield. While this will still leave the patient end ungrounded, it will create a node at the boundary where all potentials are equal,

Figure 10-7. Cable Model of Patient Cable

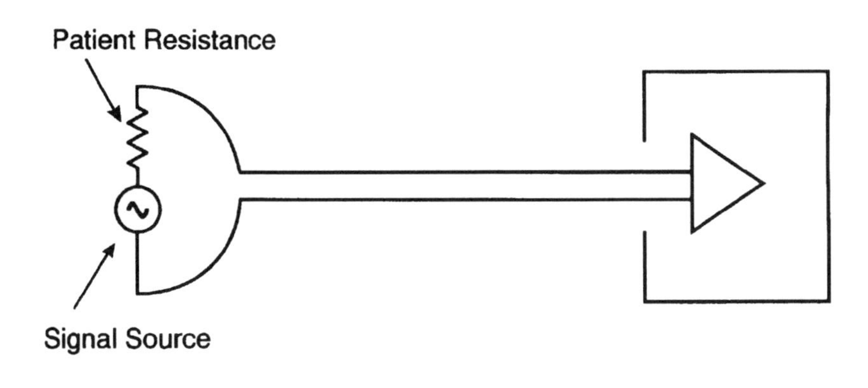

eliminating DM noise. Common mode noise will still be on both the shield and signal lines at this point, however, and this may still cause a problem at the equipment end, depending on how the shield is terminated. If the shield is hard terminated to the enclosure, then this practice will not work. If the shield is terminated as per the next item, then it will work.

- At the equipment end the preferable termination is to the enclosure. But if this is not permitted, then the next choice is to terminate to an internal shield. If a complete internal shield is possible, and the internal shield can safely be isolated from the enclosure, then normal terminations apply. Alternately, a shield local to isolated ground might be used, and the cable shield would be terminated to this ground. This protects the isolated section, which is generally the most sensitive, but still leaves the bridge to circuit ground a problem. Usually, this is easier to cope with, however.
- If neither of these possibilities exist, then all efforts must be taken to block RF from getting from the cable (including shield) to the isolated circuit area. This means high impedance filters, which must be very carefully designed. As signal line to ground capacitance is limited, the filter needs to be very high impedance, which is difficult to achieve and to maintain. Extreme measures are needed to ensure that (1) the filter is actually functional at all relevent frequencies

(watch for high frequency resonances) and (2) the components are placed to minimize electric field coupling.

Cable Shielding

After you have sealed all the connector joints, then you can worry about shield leakage itself. We will soon see, shield leakage is a concern at frequencies above about 10 MHz. Below this frequency most shields, even loose braids, perform pretty well. At frequencies above 10 MHz, however, it really pays to invest in high quality shielding for the cable covering.

Cable researchers and vendors have come up with a useful way to predict cable shield leakage. The parameter is known as the "transfer impedance," and it measures and predicts the amount of longitudinal voltage on one side of the shield as a function of current on the other side. The unit of measure is ohms/meter; the lower the number, the better the shield.

Figure 10-8 shows some typical transfer impedances for different materials. Although these apply to coaxial cables, the information is helpful to understanding all cable shield performances. Note that below 1 MHz the transfer impedances are all grouped together; above 1 MHz the braids begin to get worse, while solid materials get better. The latter is due to effects of multiple skin depths. It is apparent that above about 10 MHz, it pays to consider double braids or at least braids with high "optical coverage."

There are also sheathed Mylar® film shields, which afford lower RFI leakage. Individually, they are pretty good shields, as long as they stay intact. But they are very vulnerable to rupturing if kinked or, in the case of coiled cables, stretched to the limit. They were really not intended for use with exposed cables, but they are widely used in this fashion, anyway. Double-shielded braid and Mylar® film are very effective, and mitigate the rupture problem. Sheathing is very difficult to terminate, however. The drain wire is completely unacceptable as a termination. Worse, yet, is to run the drain wire through a connector into the circuit board.

Having said that about shields, the fact is that most high frequency cable problems are due to poor connectors. Do not bother fixing the cable shielding until you have fixed the connectors. Otherwise you are just wasting time and money. Often your cable shielding demands are not real high, and some leakage at high frequencies may not be fatal. But it is best to have some high quality cable selected and available for use if needed. You really do not want

Figure 10-8. Cable Shield Transfer Impedance

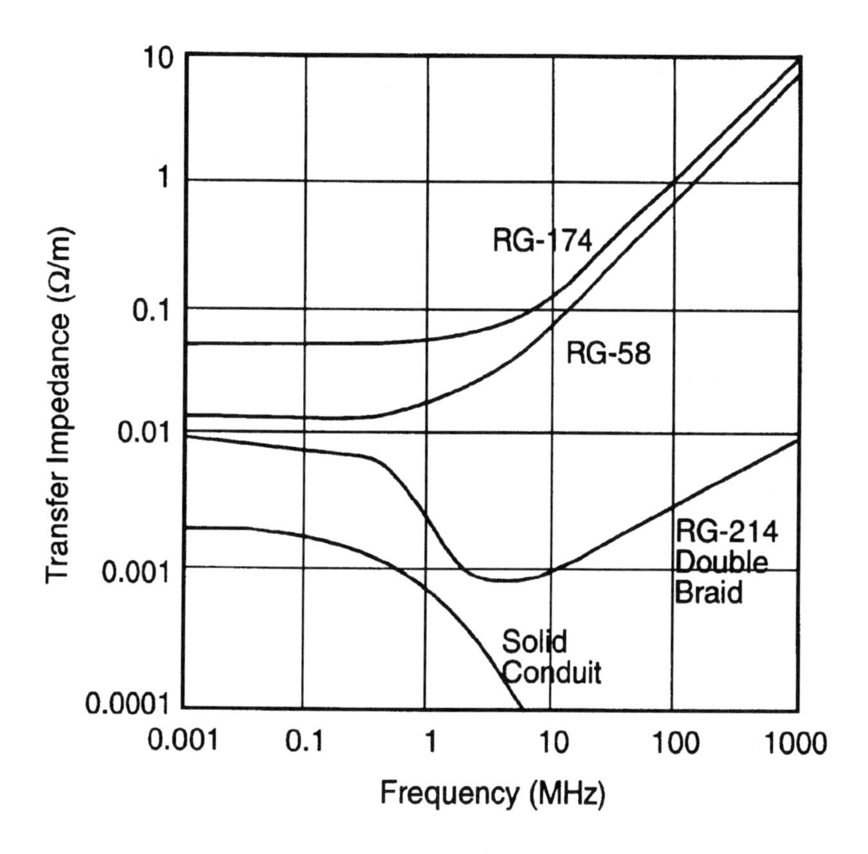

to find out at a late time in the design stage that a larger diameter cable is needed and it will not fit.

CABLE DESIGN RECOMMENDATIONS

Now that we have looked at how cables work (and how they fail), let us turn to some design guidelines for cables. First, determine the cable frequency range. As we have already seen, frequency is a critical parameter, so you must decide over what frequencies the cable must perform. And remember to consider both the operational frequency (required bandwidth of the signals or power) and the threat frequencies. Just because a cable is only intended for low frequencies, do not assume that high frequency threats will not affect it. Many cable problems are caused when low frequency cables are first subjected to high frequency such as RF or ESD.

You will also need to determine if the cable is electrically short or long, which will affect your grounding strategy. You should also determine circuit impedances for the same reason. As we have seen, even at low frequencies, high and low impedance circuits require different approaches.

Next, design your cable–do not just leave it to chance. You will need to make a number of design decisions. Decide on the internal layout and wire assignments, and be sure to include adequate ground and signal returns. Consider twisting signals and returns for DM protection and crosstalk control. Decide on the type of shielding needed. You will need high quality shielding for frequencies above 10 MHz, otherwise the choice of covering is less critical. And, finally, spend some time choosing and designing the connector interfaces. Poor connectors are the prime reasons for cable shielding failures.

As the extent of the threat is not always well known, you may well ask, "Do we really need expensive shielding or metal backshell on our connectors?" Or you may not even have an option–you are stuck with existing conditions. In this case, the following advice applies: (1) If you cannot use a metal backshell, keep the pigtail termination as short and fat as feasible–if you are lucky, you may get away with it. (2) If you have a cable shield already available, use it–if you are lucky, you may get away with it. But be aware that these deviations put your design at considerable risk, and you should be prepared to change your design if needed.

Finally, pay attention to the eventual cable routing. If the cables are inside your system, watch out for hot sources, like microprocessor clocks or other fast circuits. Oftentimes, internal ribbon cables lay on these critical circuits, and carry unwanted energy all over the place, causing problems ranging from Federal Communications Commission (FCC) emission failures to mysterious intermittent internal failures. One other piece of advice–do not route internal cables near seams or slots in shielded enclosures. This sneak path for energy has caused many failures when testing for RF emissions and immunity, as well as ESD. Figure 8-10 (on page 165) shows an example of this potentially disastrous problem.

If you are designing the equipment, you may not have much control on how external cables are routed. Nevertheless, you may want to provide some recommendations to installation personnel to minimize field problems later. The primary guide here is common sense. Keep sensitive input cables away from noisy sources. Watch out for power lines–not only are they the source of power line electric and magnetic fields, but they also carry a lot of high frequency energy, such as spikes that can play havoc with digital circuits.

SUMMARY

The following guidelines serve as a summary with external cabling:

- Single-point grounds are appropriate for cases where the shortest wavelength of interest is more than 20× the length of the cable.
- Line-to-line crosstalk is primarily a DM interference problem, degrading signals.
- Pigtail cable terminations are not permitted above audio frequencies.
- Different shielding needs exist for low frequency electric fields, low frequency magnetic fields, and high frequency fields. Assess your needs before taking an approach.
- Braid starts to become leaky above 10 MHz.

11

SPECIAL EMI PROBLEMS IN MEDICAL ELECTRONICS

Let us take a look at some interference sources found in medical environments and how to cope with them. In medical electronics there is a diversity of high amplitude sources of wide frequency range in proximity to sensitive receiving devices. Contrast this with most commercial or residential electronics (a personal computer is a good example), where there are no internal devices that pose a major self-compatibility threat and external threats are usually not severe enough to cause a problem.

One of the notable aspects of medical interference generators is that the source is often closely associated with the patient and, hence, cannot be fully shielded. Full shielding would require placing the patient and equipment operators in a shielded room–a viable, but expensive and sometimes necessary, choice.

HIGH ENERGY INTERFERENCE SOURCES

There are many high energy sources found in a medical environment. The most notorious is the electrosurgical unit (ESU), which generates fierce broad band noise from about 100 kHz to as high as 100 MHz. Typical ESUs generate a wave in the 500 kHz to 1 MHz range, but this wave is modulated at a low frequency for certain operating modes–thus the broad band noise. The principle interference problems are currents conducted out the power and magnetic fields. Low frequency magnetic fields are always difficult to shield, and require permeable materials. CRT monitors are the major, but not the only, victim of these sources. Measured electric field strengths can approach 100 V/m close to the source, enough to upset most equipment.

Diathermy has frequency components in the 15 to 30 MHz range. Interference control in this frequency range is relatively easy, as it is high enough that filtering and shielding is effective and inexpensive, while being low enough that shield gasketing is probably not needed. Measured field strengths are up to about 1/2 V/m.

Magnetic resonance imaging, or MRI, generates a very high magnetic field in the 50–100 MHz range. This is getting into the frequency range where effective shielding requires the careful control of openings. MRI rooms are well shielded and negligible energy gets outside. Radiology is characterized by high voltage power supplies, often using thyristers, a copious source of power line interference.

Light is not generally a source of interference to electronic equipment, but some significant interference sources are used in the process of powering a laser. Unfortunately, medical device manufacturers usually buy the lasers, and have but little control over the interference the product generates. You just have to cope with the interference.

The biggest threat in lasers is the magnetic field associated with a pulsed laser. The sudden discharge may produce a 100 A pulse with a 100 μs rise time. This gives an equivalent frequency of about 3 kHz, and a *di/dt* of millions of amps per second. The resulting low frequency magnetic field requires permeable materials for effective shielding.

Some continuous lasers use one of the ISM (Industrial, Scientific, and Medical) frequencies (e.g., 27 MHz) for modulation. The Federal Communications Commission (FCC) (and other government agencies worldwide) allows unlimited radiation in the ISM frequencies. As a practical matter laser manufacturers need to limit the radiated energy, or nearby electronics would not work at all. But interference control is far from perfect, leaving plenty to impact sensitive analog circuits.

All of the above devices draw copious amounts of power while generating significant amounts of interference in the process. Adequate suppression requires high quality line filters, which are large and expensive. Thus, you should be prepared to cope with a device that is inadequately filtered.

Emergency generators are found in every hospital. While not normally a source of interference, they do need to be tested periodically. Depending on the design of the power system, cutting in these generators can cause a significant power dislocation, of a type not commonly found elsewhere. Modern UPSs are capable of transitioning to and from emergency power with scarcely a ripple in the voltage, but standby engine generator sets commonly found in a hospital will not be on-line–they are started up when power fails, and are not expected to maintain uninterrupted power.

A defibrillator generates a substantial transient, with a pulse of less than 10 ms, a peak voltage of several thousand volts, and a current in the ampere range. Any equipment that may be connected to the patient at this time must withstand this transient. Any equipment operating nearby may also be affected.

ELECTROMECHANICAL DEVICES

The high energy sources cited above threaten both internal and external electronics, but much more prevalent is the case where interference originates from relatively low energy sources, which is usually a problem only for internal electronics. These sources include electric motors (e.g., blowers and stepper motors) and controls (e.g., relays and solenoids). These sources are common, but are often ignored until a problem is uncovered.

Electric motors are the source of a variety of interference signatures:

- Heavy starting loads, which cannot be effectively filtered, and must be handled by the power supply or AC line power
- Inductive turn-off transients, with peaks up to several thousand volts
- Broad band brush noise from DC motors
- Stepper motor and variable frequency drivers, which deliver a fast edge rate high current pulse

Relays and solenoid windings provide an inductive spike similar to the motor at cutoff time.

RADIO SOURCES

Radio sources abound in a medical environment. The obvious ones are television and radio, some of which are inevitably in proximity to some of your equipment. But these are not the primary sources. Close in are land mobile units, with a base unit on top of the hospital and mobile units in the emergency wing of the hospital. Handheld radios, such as used by maintenance personnel, and cellular phones are much more of a problem, as they are often in close proximity to medical equipment. Patient telemetry is a lesser, but real, threat. Awareness of this problem has resulted in hospitals banning the use of such devices in certain parts of the hospital, but this does not control the clinic, which is often in an uncontrolled facility, nor does it control the residential environment. Further, cellular phones will still transmit periodically, unless they are turned off, even if the phone is not in use.

Radar sends out a very strong burst that will be a problem in certain locations, but this is not too common. Another possibility is the hospital's practice of renting out their rooftop to transmitting devices.

RECEIVING DEVICES

We have sensitive instruments operating along side these high energy threats. While we cannot identify all of them, we can cite some representative cases.

ECG, EMG, and EEG

ECGs, EMGs, and EEGs include low frequency, low level analog amplifiers. Typically, an ECG has a 1 mV sensitivity at a bandwidth of 50 Hz. An EMG has a 100 μV sensitivity at 3 kHz and an EEG has a 50 μV sensitivity at 100 Hz. Note that all of these have a band-pass in the power line frequency range, making filtering a difficult proposition. The ECG, while least sensitive, is most often exposed to a hostile environment–they are used everywhere in a hospital. These sensitive front ends are most sensitive to RFI.

Ultrasound

Typically, ultrasound units generate high frequency (MHz range) acoustic waves, and are looking for a very low level echo, typically in the 100 nV range. These devices are primarily a threat to themselves, being exposed to their own signals.

Respiratory Devices

Apnea monitors are characterized by a low level signal being sensed by sensitive amplifiers. As with the ECG, they are quite sensitive to RFI, and, worst, they are usually found in an uncontrolled residential environment.

Pacemakers

Pacemakers are obviously exposed to a completely uncontrolled environment and, even worse, may be subjected to a defibrillator jolt. Fortunately, these devices are very compartmentalized, and are capable of being immunized from even severe threats.

Telemetry

Increasingly found in the hospital environment, telemetry operates on either of three bands (approximately 160 MHz, 460 MHz, and 900 MHz). These are not usually threatened with radio sources, but they are threatened with loss of signal. Expect to see more of these issues in the future as this technology proliferates.

LEAKAGE CURRENT LIMITATIONS

As mentioned in previous chapters, leakage current limitations pose a significant impediment to EMC design. For patient-connected devices the limitation is 20 μA, which is measured as in Figure 11-1. This results in a total filter capacitance to ground of less than 500 pF, and with inevitable parasitic capacitances, the capacitance available for filtering may be only 100 pF. Chassis leakage limits (nonpatient connected) are either 100 or 500 μA, depending on the category–still not a lot.

Figure 11-1. Measuring Leakage Currents

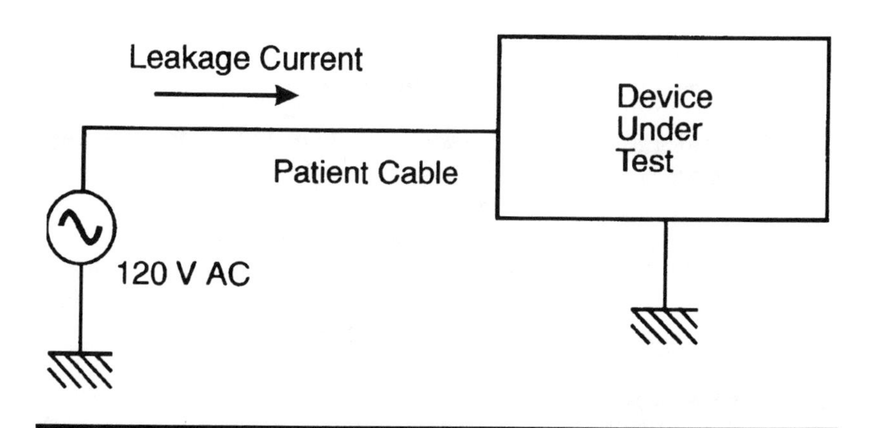

COPING WITH INTERNAL THREATS

How do you cope with internal threats? Start by assuming that all high energy sources are a threat, and look for methods of dealing with them.

Load

You can start by isolating the power for the various loads. Generally, loads can be divided into four categories, RF generators, electromechanical (50–60 Hz), digital, and low level analog. You might as well start by providing separate power to these loads, preferably by using separate supplies.

Electromechanical and RF generators usually do not need high quality AC power (they are usually the source rather than the recipient of interference). Provide them with a separate power source, one with adequate reserves to cope with starting transients.

Interference

Power line interference as generated by power lasers or any other high power load can be quite difficult to remove from a system (these devices often use thyristers for regulation, generating lots of interference). The best solution is to have a completely separate power feed for the laser. It is fairly common for high power systems to draw 480 V AC power. Unfortunately, this is usually tapped down to get 120 V AC to feed electronics. Electronics will be far happier drawing power from a separate 120 V AC source, (Figure 11-2), which provides pretty good conductive isolation from high energy interference generators.

On the receiving end of interference is the analog device, which is often tasked to sense voltages in the microvolt and sometimes even in the nanovolt range. These devices will be susceptible to both internal and external RF sources, even if the offending frequency source is much higher than the band-pass of the analog device. If the analog circuits are not highly sensitive, then it may be sufficient to provide filter isolation between analog and digital circuits. But by the time you are in the low microvolt range, you clearly need to take substantial efforts to obtain isolation. Cable shielding or optical isolation techniques may be advised.

Magnetic Fields

Where intrasystem magnetic fields are a problem, partition your system so that the offending field (for a pulsed laser, this will be in the

Figure 11-2. Power Source Separation

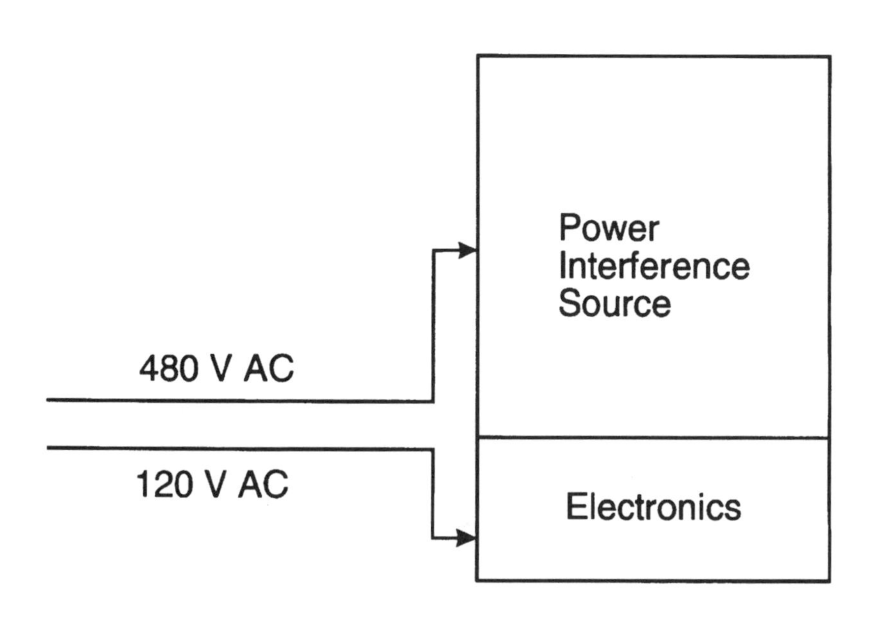

pulse current path) is physically as far as possible from your sensitive circuits. Magnetic field shielding is most effective when placed as far from the source as possible and this often means shielding the receptor circuits. Good internal cable routing practices are important here (see chapter 8), keeping the loop areas as small as possible by routing the cables over the ground surface.

The huge low frequency magnetic fields generated by pulsed lasers are very difficult to contain. Use permeable materials for shielding. Mild steel is preferred for most cases, as exotic high permeability materials are hard to work with and saturate easily. Closed magnetic paths are essential. Welded seams are the best, but overlapping seams with frequent fasteners may be sufficient. By the way, grounding plays no part in magnetic field shielding effectiveness.

The RF modulating source in continuous lasers is probably moderately well shielded, but it is unlikely that the shielding is adequate for nearby low level analog signals, especially those that are on cables. You will probably be better off shielding and filtering your sensitive circuits, as described in previous chapters: modifying someone else's equipment is usually an unpalatable choice.

Motors

The inductive kick from motor and relay cutoffs can be suppressed with a diode snubber immediately (Figure 11-3) at the load, thus providing a path for the inductive current. This will not work for bipolar devices (AC drive, reversible DC, or stepper motors), where the current may be either positive or negative. In such a case an RC circuit is a compromise solution (not a very good one, though). Select the time constant to provide an effective shunt for high frequency spikes, while being transparent to normal load current requirements. The snubber should be placed as close to the load as possible.

Broad band brush noise from a DC motor is usually adequately filtered with a capacitor (1nF to 1μF) across the power feed, directly at the motor leads.

Figure 11-3. Snubbers for Inductive Noise Suppression

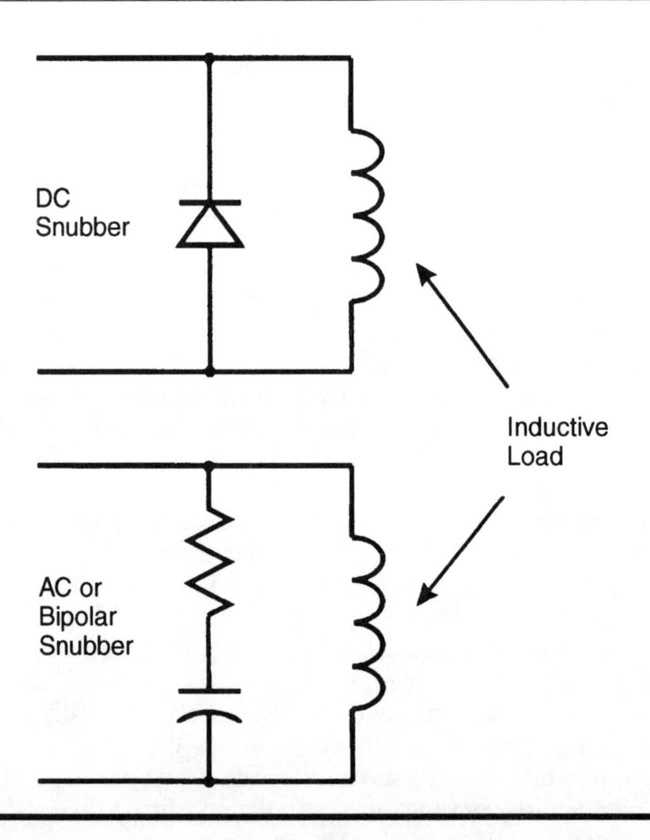

Stepper motor and variable frequency drives need to be suppressed. Designers like using fast power switching devices to minimize device heating, but they need to be aware that these fast devices are a major source of local interference. The savings in switch size may well be offset by the additional filtering needed to contain the switching noise. The best place to stop the switching noise is at the source, and this demands filters on both the power line and on the stepper current output. Steppers and variable frequency drives (VFD) drivers are low impedance output, and do not take kindly to shunt capacitive filtering. Use inductors in series with these outputs before any capacitance. Your filter manufacturer has filters designed expressly for this purpose.

Regardless of suppression used, interference effects can be minimized by parallel runs of the power and return, and signal and return, preferably by twisting and by routing next to a continuous ground surface. This will minimize coupling to other nearby circuits.

COPING WITH EXTERNAL THREATS

The best defense for external threats is distance. Unfortunately, this is not always possible. Extremely sensitive or noisy equipment may need to be installed in a shielded room. These can be bought or built on-site, but careful installation is needed to maintain shield integrity. We find that construction people need to be closely monitored, as most are not yet trained in techniques of handling electromagnetic interference (EMI)—they will often ignore specifications, mostly because they do not understand the significance of them.

A word of caution on most screen rooms: They will not shield low frequency magnetic fields. If you need magnetic shielding, be sure to identify the need and the frequency when selecting the shield room.

Most threats can be addressed in a less drastic fashion, however. Conducted interference on the power line is usually avoided by using dedicated circuits (separate power and neutral), but if that is insufficient or impossible, then power line filters may be needed or supplemented. Power conditioners or UPS see widespread application in such cases.

Interference can often be suppressed simply by use of a ground plane. For safety reasons equipment is well grounded, but safety grounds do not meet the criteria for good RF grounds. Grounds were discussed in chapter 6.

Open loop areas between the cabling and the ground surface intercept much radiated energy. Conducted interference due to ground noise between interconnected devices also causes problems, primarily at power line frequencies and low order harmonics.

For best results all interconnected equipment should be placed over (and connected to) a ground plane. Ground grids in a computer room approximate this ideal. Where a full plane is not possible, interconnect a wide metal strap between the equipment. Where high frequencies are present (1 MHz and above), keep the length no longer than five times the width. Of equal importance is the routing of data cables along the ground surface.

SUMMARY

When designing medical electronics, you need to evaluate the myriad of interference sources, both internal to your equipment and external. Once identified, consider FAT-ID (frequency, amplitude, time, impedance, and dimensions) to help select the appropriate remedial measures.

- Do you have internal noise sources (laser, motor, etc.)? If so, contain at source.
- Are you likely exposed to external noise sources (ESU, MRI, radio)? If so, identify methods of fixing at recipient.
- High frequency, high energy RFI threats probably need to be handled by shielding. Take care to close the seams with gasketing and minimizing openings. Also, unshielded lines need filtering.
- Low frequency magnetic fields need permeable shields.
- Power threats are a major problem in hospital environments, and need filtering or suppression to contain.
- Intercept the interference at the source if possible. If not possible, concentrate on protecting the most sensitive circuits.
- Good grounds are mandatory for successful operation of electronics, but do not expect safety grounds to serve for RFI.

12

SYSTEMS CONSIDERATIONS

Placing medical electronic equipment in a hospital or clinical environment often gives rise to equipment anomalies. Your sensitive equipment is assaulted by external interference, or perhaps you are the interference source and other equipment is affected. Even if you have designed your equipment in accordance with prevailing EMI standards, there is no guarantee that it will be compatible in its existing environment. Remember, EMC means satisfactory operation in its intended environment. It does not guarantee that all environments are compliant with standards.

When your equipment causes problems in place, then you are left with two alternatives: Fix your equipment or fix the environment (better yet, fix the environment before installation). Fixing the environment necessarily involves preparing the facility in some manner. The fix may be minimal or it may be major. Here are four scenarios:

1. Your equipment is very noisy by nature, and is apt to upset adjacent equipment. In this case we presume that all feasible possibilities to contain the interference by design have been taken, and that external methods will be required.
2. Your equipment is very sensitive by nature, and is apt to be upset by adjacent equipment. Again, we presume that all feasible steps have been taken to contain the interference by design, and that external methods will be required.

   Both of these scenarios may require a screen room or shield room, along with power line filters. These needs, being inherent in the nature of the equipment, would presumably be known in advance of installation. Accordingly, advance preparation is needed.

3. Your equipment is neither unusually sensitive nor noisy—it is a typical piece of equipment. There are numerous installations, almost all of which are operating as designed, but a few isolated installations do not perform satisfactorily. As there are few problems, it is presumed that the conditions causing the problems are peculiar to the local environment. Accordingly, your goal may be to identify a facility fix, without resorting to product design changes. You will need to understand the source of the problem before a fix can be identified.
4. Your equipment is neither unusually sensitive nor noisy by nature. It is a fairly new product and problems are surfacing with alarming frequency. You suspect there is a latent susceptibility. You want to fix the problem at the equipment level if possible, but first you need to find out the cause and, hopefully, identify a quick fix.

Thus, we have two problems, troubleshooting and site preparation. We need to understand the problems before we can decide what needs to be done. Subsequently, we can decide if the equipment is to be redesigned or if the site is to be prepared in some fashion.

IDENTIFYING EMI THREATS AT THE SITE

When an EMI problem occurs at the site, it is necessary to identify the problem before corrective action can commence. In some cases you will want to find the problem so that you can go back to your lab and work on a fix. In other cases you will want to fix the problem on the spot. In either case you need to have some idea what is causing the problem before attempting a fix. On-site problems are rarely kind enough to occur with sufficient regularity that you can observe them on the spot. If possible, you would like to be able to force the failures so that the effect of the fix can be observed on the spot. Let us look at the three key threats and see how they might be identified.

Identifying RFI Threats

The most likely sources of RFI are communications transmitters, so take a quick look around. As an even faster estimate, you can assume commercial transmitters are not a problem at greater than about 2 kilometers (a little over a mile) from the antenna, handheld radios and cellular phones at greater than about 2 meters, and mobile or fixed-base radios at greater than about 20 meters.

If you are in a hospital, look for "unintended radiators," such as lasers, cauterizers, MRI systems, or diathermy machines. If you are in an industrial environment, look for welders or RF heat-sealing equipment. If you are in a residential environment, look for citizens band (CB) or amateur radio antennas. If you are in a vehicular environment, look for both on-board and nearby mobile transmitters, as well as ignition noise.

You may need to make some field intensity measurements. For a quick check you can use a "field intensity" or "field strength" meter to determine the levels, but you will need a receiver or spectrum analyzer to determine the precise frequencies as well. If the offending frequency is known, a field strength meter will work fine. You can even make an approximation, using the following formula:

$$\boldsymbol{E} = 5.5\frac{\sqrt{P}}{d}$$

where E is the field intensity in volts/meter, P is the effective radiated power in watts, and d is the distance from the source antenna in meters. Remember, anything above about 1 volt/meter is suspect.

If you can determine the offending frequencies, you can guess at the path (<30 MHz, conducted; >30 MHz radiated to cables; >300 MHz radiated to cables, box, or circuit boards). If you can, disconnect cables and unnecessary equipment. If you find that the problem goes away when the cables are removed, you have found the unwanted antenna as well. If the problem still persists when all cables are removed, you might guess that the frequency is above 300 MHz, and that it is radiated directly to the enclosure or circuit boards.

Identifying ESD Threats

Common complaints surrounding ESD is circuit lockup. The system resets or simply hangs. How do you know it is an ESD problem rather than an RFI or power problem? Well, in many cases it will be difficult to pin things down, you must use the process of elimination.

You can look for clues, however. First, the most common symptom of ESD is that the problem surfaces in the fall, along with the heating season. Equipment that which was installed during the humid summer months will first experience low humidity in the fall.

If you do not have such a smoking gun, you need to look for other indicators. What is the humidity at the location? Drawing sparks from your fingertip or cracked and bleeding lips are dead giveaways, but usually you are looking for less obvious indications. Do people complain about low humidity or perhaps they simply have a dry cough?

Look for other indicators that might provide for an ESD source. Carpeting is a common source, with floor tile in second place. If the equipment has electromechanical devices, look for moving belts, rubber rollers, and so on.

Fortunately, ESD guns are readily available, and they provide an excellent on-line method of forcing failures. A major word of caution, however: ESD is especially prone to destroy components, so proceed with caution. You would be wise to have a supply of replacement modules available.

Indirect discharge is far more benign than direct discharge, so do it first: discharge to metal members in the immediate proximity of the equipment. Start with low voltages (typically 2 kV) and proceed to higher voltages in 2 kV increments. If indirect discharge has no effect, try direct discharge and repeat the process. As soon as you find a soft spot, carefully document vulnerable areas and associated susceptibility thresholds. Then look for a possible fix, implement it, then resume to see if improvement has been achieved.

We have repeatedly seen cases where ESD was unsuspected, but that was ultimately found to be the problem. So do not make any hasty assumptions–check them out.

Identifying Power Disturbance Threats

Common complaints relating to power disturbances are system reset or lockup. How do you know it is a power disturbance rather than ESD or RFI? Well, in many cases it will be difficult to pin things down and you must use the process of elimination.

You can look for clues, however. First, the most common symptom of power disturbances is that it occurs at specific times of the day or at particular times of the year. An operating room is most commonly used in the morning, for example, and air conditioners operate in the summer.

If you do not have such a smoking gun, you need to look for other indicators. Inspect the facility in the immediate proximity of the malfunctioning equipment. In many cases the disturbance will be from equipment powered from the same circuit. If the equipment is installed in a hospital, there will usually be separate circuits for critical or noisy equipment. Separate circuits are less likely in a clinic; in a residence any wiring condition can exist. Note: "Separate circuits" specifically mean separate neutrals as well as separate "hot" lines.

Power line monitors will detect most power line disturbances, including transients; however, they are limited to about 2 MHz bandwidth. Electrical fast transients (EFTs) are too fast for power monitors, but can be detected by high speed digital storage oscilloscopes.

Power disturbance generators are readily available for testing in the lab, either for compliance testing or for failure forcers. These are usually not feasible for use in the field.

If you do not have such equipment available, then you might try a "chattering relay." A chattering relay is one that is wired across the line with the windings in series with the normally closed relay contacts. The relay does not know whether it is on or off, so it chatters, sending out copious amounts of interference. The chattering relay test may be too severe, so you may need to modify it a bit to reduce the amplitude.

SITE PREPARATION

Do not let this heading intimidate you. Site preparation may be quite simple or quite elaborate, depending on the situation. But, in most cases, preparation is simple.

There are three issues of interest when preparing a site: grounding, power quality, and shielding. Of the three, shielding is usually extensive, whereas grounding and power quality may be quite simple.

Grounding was discussed in chapter 6. As mentioned, grounds mean different things to different people. We make no attempt to define safety and lightning grounding needs–but we may make use of these grounds or supplement them. The National Electrical Code (NEC) goes to great lengths to define acceptable practices: NEVER violate it for any reason–it makes good sense.

EMI may occur over a wide frequency range, from 60 Hz on up into the GHz range, and currents may be amps or microamps. The interference may be transient or it may be quasi-continuous. In all cases, however, the fundamental problem is the same–to provide a low impedance path back to the source of the interference, or at least to minimize voltage differences between interoperating equipment. Note that we are not talking about an earth ground, but rather a ground reference local to the affected equipment. And also note the extremely wide range over which an EMI ground may be tasked to operate.

Grounding Concepts Made Simple

We can really simplify the grounding concepts to a few key points.

1. Keep ground impedances low. How low? This depends on the nature of the interference. If you are trying to keep two high frequency equipments at the same potential, then a low impedance ground may be needed in the MHz range. If you are looking at power line–related interference (60 Hz or low order harmonics, plus higher frequency switching spikes), then a low impedance ground good up to 10 kHz will be adequate.
2. Divert ground currents away from sensitive equipment. This is simply recognizing that ground noise may well be such that it cannot readily be kept at a low enough level to avoid problems. As an example, facility ground noise from one plug to another can easily be several volts, and it takes some pretty stiff grounding to reduce it. Accordingly, we look for ways for avoiding a confrontation. This generally is accomplished with the use of a single-point ground. Single-point grounds are not important when discussing equipment issues, but are usually quite significant when dealing with facility issues.

Low Impedance Grounds

As mentioned, we cannot really impact facility ground currents with any degree of success, but we can establish local ground references that are fairly stable.

Let us return to the impedance of a wire. In a previous chapter we gave an approximation of 8 nH/cm of wire, a value that varies only slowly with wire diameter. Actually, the function is superlinear, so the approximation is less good for longer wires. For facility distances and wire gauges we use the approximation L = one microhenry/meter, which is reasonably valid for heavier wire gauges and longer wires. Again, inductance varies only slowly with wire diameter.

For example, two pieces of equipment placed 2 meters apart are threatened with 1 MHz ground noise. What would be the impedance between the two boxes if we connected them with a length of heavy gauge wire?

$$L = 1 \times 2 = 2\ \mu\text{H}$$

$$Z = 2\pi \times f \times L = 6.28 \times 1 \times 10^{6} \times 1 \times 10^{-6} = 12\ \Omega$$

If noise currents were measured to be 100 mA, the voltage would be

$$V = IZ = 100e^{-3} \times 12 = 120\ \text{mV}$$

This is at the threshold of being enough to upset digital signal levels.

This situation is helped only marginally by installing a strap. A 2 cm wide strap, for example, would reduce the inductance only about 25 percent, certainly noticeable, but not enough to provide significant relief.

The message is that if you really want low impedance interconnect at radio frequencies, use a wide ground strap or a ground grid as shown in Figure 12-1. As currents are not significant (facility ground carries the heavy low frequency currents), the thickness of the ground strap is not a significant factor.

Single-Point Grounds

Single-point grounds are established to prevent sensitive equipment from sharing ground paths with noisy equipment. But single-point grounds do not make sense at high frequencies. We use the criteria that multipoint grounds are appropriate at lengths greater than 1/20 of a wavelength of the highest threat frequency. This is where the wire starts to look like a fairly efficient antenna. It is also roughly the point where single-point grounds simply do not exist, in spite of best intentions. Figure 6-7 shows the single-point concept–two boxes interconnected by a single-point ground (inductive to the tune of 1 μH/m) and mounted on a surface. Lacking any other values, we find that an enclosure capacitance to ground of 100 pF is fairly typical.

For example, what are the relative impedances at 30 MHz of two enclosures placed 3 meters apart and grounded to a common, single-point ground midway between? Assume 100 pF per enclosure.

$$L = 1\ \mu\text{H/m} = 3\ \mu\text{H for 3 m}$$

$$Z_L = 2\pi \times 30 \times 10^{6} \times 3 \times 10^{-6} = 560\ \text{ohms}$$

$$C = 100\ \text{pF each} = 50\ \text{pF in parallel}$$

$$Z_C = 1/(2\pi \times 50 \times 10^{-12} \times 30 \times 10^{6}) = 110\ \text{ohms}$$

Figure 12-1. Low Impedance Ground Grid

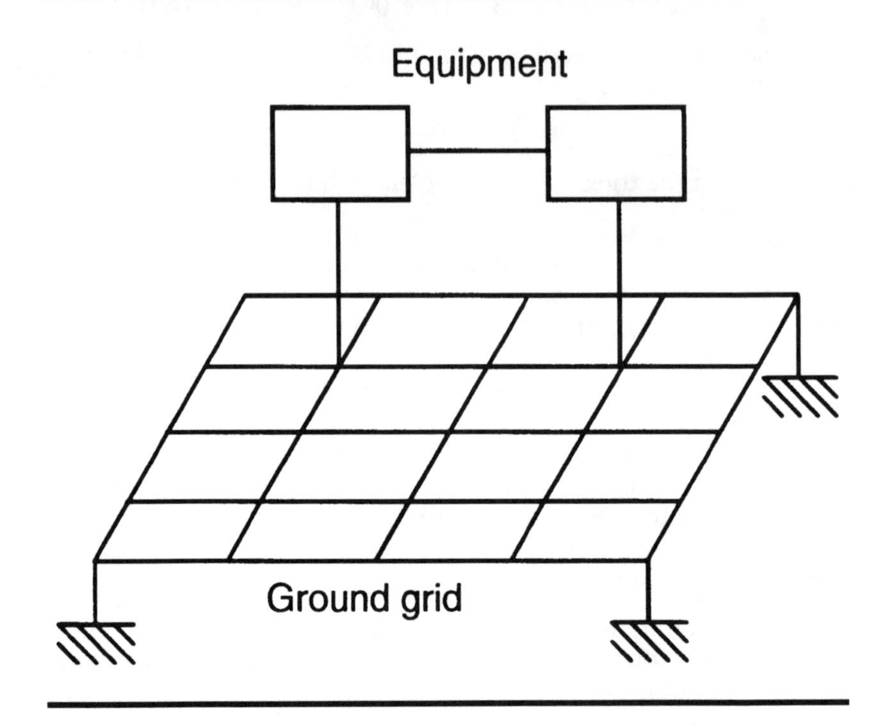

Thus, we see that at 30 MHz the capacitive path is significantly lower in impedance than the single-point ground path at 30 MHz, and the condition deteriorates at higher frequencies.

So we see that single-point grounds do not exist for higher frequencies. But for low frequencies (say audio frequencies), the single-point ground is very appropriate. As most power disturbances fall into this category, it is appropriate to use them.

Power Quality

Power disturbances have been defined as a deviation from the ideal sine wave (50 or 60 Hz) that we would like. The two most common problems are voltage sag and the spike (EFT), as mentioned earlier. Both are a common problem in a facility. What can we do about them?

We can start with standard wiring practices as per Figure 12-2. In case one, two devices share the circuit. Any noise currents

Figure 12-2. Wiring Practices

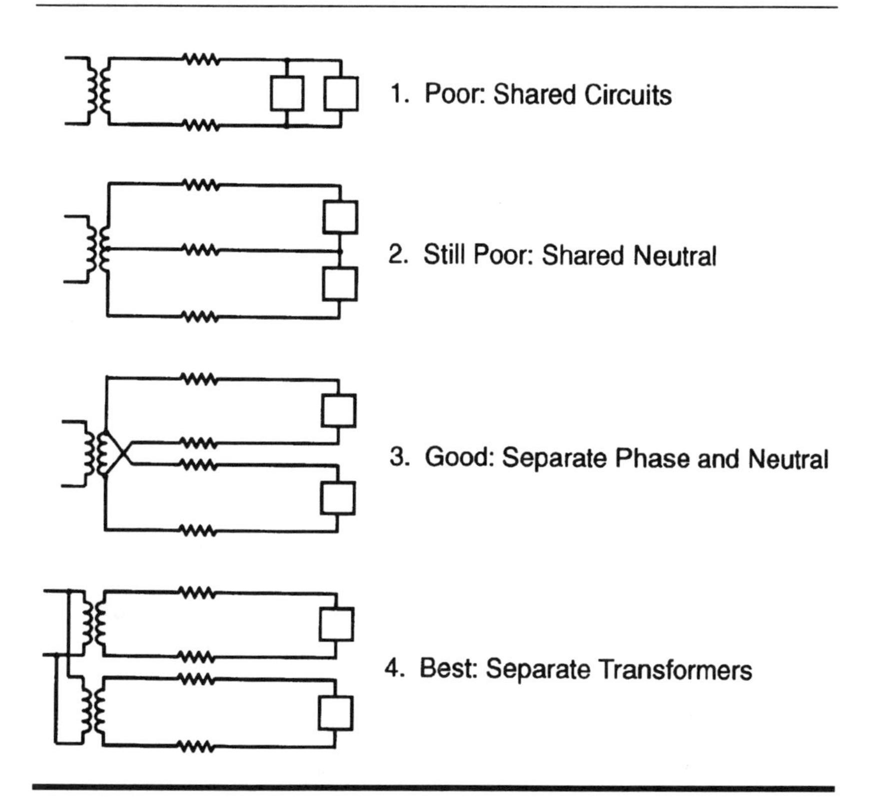

generated by one will divide between the inductive path back to the source and the other load. This is the most common victim of an EFT. We see it time after time. Real common problems are plugging your computer into a circuit shared by a refrigerator or a small copier. Our rules are simple: Power your electronics from a separate circuit.

In case two we have divided the circuit, but failed to separate the neutrals. This is a common practice where three phase plus neutral is routed to a junction box, then divided into three branch circuits, all on the same neutral. This solves half the problem mentioned above.

In case three we have separate circuits all the way back to the transformer. Now the offending noise has to travel all the way down to the transformer, encounter the low impedance of the transformer, then travel all the way back to another load. This is a pretty good solution, and is the one usually recommended.

In case four we have circuits powered from separate transformers. This is a common practice in industrial environments where heavy equipment loads are powered from a large transformer (often 480 V) while office equipment is powered from a smaller transformer feeding 120/240 V. This provides very good isolation between the two loads.

Power Conditioners

When power quality problems are suspected, power conditioners are often employed. In many cases it works; sometimes it does not work. When it does not work, you have addressed the wrong problem. Power conditioners perform several vital functions, depending on the type:

- Spike and high frequency suppression
- Regulation
- Ground/neutral restoration

The first two are truly power quality problems, and they are handled quite well by power conditioners. The third is really a ground noise problem, and is also handled quite well by some power conditioners PROVIDED they are installed correctly.

The ground noise problem is related to noise originating elsewhere in the facility, and will occur regardless of efforts to provide clean power (either by separate circuits or by power conditioning). The key is employment of a power transformer, which reestablishes the neutral/ground connection. In order to clean up ground noise, two conditions are necessary:

1. Ground must be reestablished. This is always done when the power conditioner has a power transformer, but is never done when the power conditioner uses an autotransformer. The neutral/ground noise will remain unchanged if an autotransformer is used. Note that this function is independent of power quality on the power feed, and can be cured by using a power transformer.
2. Single-point grounding must be employed. This means that from the power conditioner (or power transformer) to the loads, a single-point ground is provided (Figure 12-3). The loads must have no other ground path, which may occur either because of chassis grounding or by signal cable ground paths. If single-point grounding is not used, ground noise will pass through your equipment even if a power conditioner is used.

Figure 12-3. Power Conditioners and Grounding

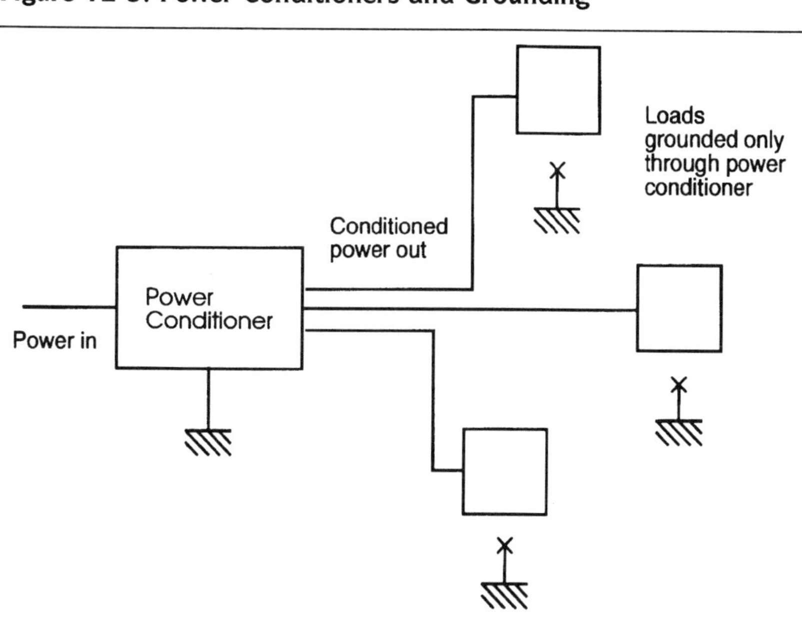

Shielding

In most cases facility shielding will not be needed. But three conditions may give rise to the need.

1. Your equipment is noisy and must be shielded to protect nearby equipment. If this is the case, you are already cognizant of the problem and solution.

2. Your equipment is very sensitive and must be shielded to protect it from nearby equipment. If this is the case, you are also cognizant of the problem and solution.

In both of these cases, the solution is generally a screen room or shield room, selected to provide the necessary shielding for the problem frequency spectrum and attenuation needed. There a number of vendors that specialize in designing, building, and installing such facilities. A word of caution, however; correct installation is mandatory, or the room will not perform as needed. Those experienced in such installations know what to do, but most construction crews do not–they will inevitably misinstall.

3. You have ordinary equipment and have an adverse environment, which may be known ahead of time or may be

uncovered after installation is completed. Such an environment is most often a result of installation near a radio source–perhaps a nearby TV tower, a facility radio system, or operating room. This type of problem will be discussed below.

Architectural Shielding

There are a number of options with architectural shielding, all of which assume you are intending to shield an entire room. The type to use depends on the frequency of the threat and the magnitude, as well as the state of the facility at the time of implementation.

Shielding needs generally come in two flavors–low frequency magnetic fields and high frequency electromagnetic fields. Low frequency magnetic fields need permeable materials (either steel or mumetals), and plenty of it. Thickness is a key factor; therefore any thin shielding material will not work. You need heavy gauge sheet metal as a minimum. This is a very expensive proposition at the facility level, and should be handled at the equipment level if at all possible. High frequency electromagnetic field shielding needs are much more common, and can be handled with a thin layer of conductive material.

It is necessary to decide how much shielding effectiveness is needed, and this can only be done accurately by knowing how much radio energy is present and how much your equipment will withstand. An RF survey will determine the offending field strength. If this is not possible, but the source power, frequency, and distance is known, a estimate can be made. Electric field strength can be approximated by

$$E = 5.5\frac{\sqrt{P}}{d}$$

where P is effective radiated power in watts, R is distance from the source in meters, and E is field strength in volts/meter.

For example, assume a 35-watt transmitter from an emergency vehicle parked 10 meters away.

$$E = 5.5\sqrt{35/10} = \text{ 3.3 volts/meter}$$

Your equipment needs can be established by test. For those who test to IEC 1000-4-3, most will use the 3 volt/meter requirement. Note that by the above example, your equipment would be at some risk to the stated threat. In practice, such estimates are never really reliable, as the threat amplitude may not be that high (or it may be higher) and the receiving device probably will not fail at precisely 3 V/m at that specific frequency. Our own recommendations would vary depending on the criticality of the mission and on the present state of the facility.

- If the mission is critical and the room is not yet built, shielding the room from the existing threat and any future threats may be advisable. Be aware, however, that such an effort will cost tens of thousands of dollars.
- If the equipment is installed and operates with occasional anomalies, then low performance shielding will usually suffice.
- If the threat is measured or projected to be above 5 V/m, then some shielding is advised. Generally, a 30 dB shield will suffice.
- If the threat is measured or projected to be above 20 V/m, then a higher quality shield is advised, say 60 dB. This condition is quite rare, and will not occur unless you are very close to a high energy noise generator.

You can estimate the shielding needs by the relationship

$$SE = 20 \log (E_s/E_e)$$

where SE is shielding effectiveness in dB, E_S is the measured or calculated field strength of the offending source in V/m, and E_e is the known or projected field strength that the equipment can tolerate, also in V/m.

If we assume the equipment will withstand 3 V/m, we can compute the following table:

Es (V/m)	SE (dB)
3	0
30	20
300	40

As can be seen, a shielding effectiveness of 20 dB will suffice for most facility shielding needs, and this is readily achieved with contemporary materials.

Shielding Materials

For electromagnetic field shielding thin conductive coatings are adequate, either with conductive paints, conductive films (such as employed in conductive wallpaper), and conductive mats (which are employed as an underlayment). The effectiveness of the shielding is dependent on the conductivity of the material. In most cases the vendor will have data regarding the shielding effectiveness of the media as a function of frequency. If not, but you do have data on the conductivity (usually given in ohms/square), you can make a crude estimate of the shielding effectiveness using the following relationship:

$$SE\ (\text{dB}) = 20 \log Z_w/4Z_b$$

where Z_w is the impedance of the incident wave in ohms and Z_b is the impedance of the barrier in ohms/square. In the far field (which is usually the case), Z_w is 377 ohms. Shielding effectiveness of a thin barrier is quite high for low frequencies, then decreases at higher frequencies, due to skin depth considerations, then finally increases again at highest frequencies as absorption (discussed in chapter 8) becomes significant. The upshot is that estimating the shielding effectiveness is chancy, and you should give yourself a 20 dB margin.

> For example what is the shielding effectiveness of a coating with Z_b of 0.1 ohm/square?
>
> $$SE = 20 \log (377/0.1) = 71\ \text{dB}$$
>
> Allowing 20 dB decrease for worst-case frequencies, your SE is about 50 dB, which is sufficient for most shielding applications.

Thus, our rule for barrier impedance is that 0.1 ohm/square is about as high a resistance as you should go. Carbon-based coatings are too high for adequate shielding. Transparent coatings are quite thin, and typically have barrier impedances of 50 ohm/square, which is not very good. Nevertheless, it is better than nothing, and is often used for windows, where transparency is desired. If the window cross section is not excessive, coated windows will not seriously degrade the overall shielding effectiveness.

MAINTAINING SHIELDING EFFECTIVENESS

As was discussed in chapter 8, the real problem with shielding effectiveness is not the material, but with seams and penetrations. Let us continue this concept to the facility level.

Seams

Installation of the conductive layer must be lapped to ensure continuity between adjacent panels, and between ceiling, walls, and floor. Remember our 1/20 of a wavelength criteria: The seams must be closed at least every 1/20 of a wavelength of the threat frequency. If the threat is a 150 MHz transmitter, the wavelength is about 2 meters and your seams must be no longer than 10 cm, or about 4 inches. This criteria will give you a 20 dB shield, regardless of the quality of the material. You should, on the average, have complete lapping, leaving gaps only in rare circumstances. Your room does need six-sided shielding, by the way–radio waves come up from below or down from above as well as from the side.

Openings

Once the seams have been sealed, you have openings remaining–windows, doors, and ventilation. The needs here depend on the overall shielding requirements. Windows can be closed using transparent coatings as mentioned above, recognizing that the conductivity is typically not real good. This film must be conductively mated to the wall shielding, at least in the corners (this will violate our 1/20 of a wavelength rule), and preferably more often. It does no good to have a conductive window without mating it to the surrounding shield. Doors would need to be conductive to complete the shield, including conductive gasketing on the sides, top, and bottom. Ventilation paths can be closed by a simple metal screen.

Is all this needed? This depends on the situation. Simply putting in conductive wallpaper will provide a measure of shielding, and at relatively modest expense. Conductive coating of windows is not expensive, but mating it to its surroundings is a problem. Metal screens over ventilation openings are, of course, no problem. If the shielding needs are known to be significant, then these rules must be followed. But if the problem is not well known, or if it is estimated to be modest, our advice is to install the shielding in the walls and ignore the windows and doors for the moment. In moderate cases the shielding will be enough. If it is not enough, shielded windows and doors can be installed without major retrofit.

Penetrations

Finally, the shielded room is subject to conductive penetrations by signal lines, power lines, as well as facility penetrations such as building steel, water pipes, and air plenums. These need to be terminated, as much as possible, as discussed in chapter 10 on cabling.

Specifically, all metallic penetrations need to be terminated to the shield–meaning a ground strap from the shield to each conductive penetrating member, including cable shields. Conduit and data cables not destined to the room should be routed around. Thermostat wires should be routed outside the shield, only penetrating out to the thermostat itself. Signal lines and all power lines cannot be terminated to the shield, and they may need filtering at the shield boundary. Again, modest shielding needs may allow some relaxation of these requirements. Power line filtering can be expensive, especially where there are a number of lines to filter (all power lines are included, even those used for lighting).

TROUBLESHOOTING IN THE FIELD

Suppose you have a problem in the field and you would like to find a fix while there. Perhaps you are looking for a facility fix. Or perhaps you have been unable to duplicate your problem in your own plant, and you are trying to identify a fix on the spot. What do you do?

Well, there are some things you can work with at the enclosure level. Understanding that a field patch is not going to be a satisfactory permanent fix, but it can be very beneficial for identifying what needs to be done. So let us look at the steps you would take in the process. You will need some useful components for fixing in the field:

- *Shielding materials.* Most commonly, this is simply aluminum foil bought from the store, although heavier industrial grades are larger in size and more durable, if available. Copper foil is better, but much more expensive, less readily available, and mostly unnecessary for field work. If you are concerned about low frequency magnetic fields, you will need some permeable shielding. You can buy rolls of mumetal, which can be wrapped around the device in question, but working with magnetic shields is not typically a field endeavor.
- *Conductive tape.* Such tapes are most often copper and aluminum, and they may or not have a conductive adhesive side. If the adhesive is not conductive, it is not very useful for troubleshooting. The tape can be used to close shield openings or to provide a temporary ground strap for short lengths.
- *Ground straps or heavy gauge wire.* This would include a braided strap, available from truck or implement suppliers,

and heavy wire available from electrical suppliers. These are useful for establishing low frequency ground paths. In general, flat braid is not materially better than round conductors for such lengths, but some people feel more comfortable using it, and it will do no harm.

- *Power line filters.* Generally, these are included in existing designs, and additional filtering is not commonly needed. Nevertheless, it is nice to have some commercial filters available. In particular, low leakage filters commonly found in medical devices are often not up to the task. As long as you are not in an operational environment, you can temporarily increase the leakage with a filter to find out if it helps.
- *Signal line filters.* There are a number of plug-in filters available for data ports, such as RS-232 lines. They can be quickly installed on-site.
- *Bulk ferrites.* These are split ferrites suitable for clamping on power and data cables for the purpose of suppressing common mode (CM) interference. They are available in various sizes and shapes for clamping around ribbon cables, either internally or externally.

What to Do?

Troubleshooting is often a case of trial and error, but there are some common threads to follow:

- If there is more than one operational mode in the system, cycle through to identify if there is a particular mode where the problem is dominant. In many cases you will find that there is a specific mode causing the problem. Concentrate on this mode.
- In any troubleshooting situation it is best to start by reducing your problem to as few elements as possible. Any cables and attached equipment not essential to the operation or to the equipment (at least in its failure mode) should be removed to ensure they are not contributing factors.
- If possible, try to force the failure. This may mean using an ESD gun, a chattering relay plugged into the same outlet, or a handheld radio. If power is suspect, try cycling nearby equipment to see what happens. If you cannot force the failure, and the failure occurs infrequently, your probability of short-term success is low.

- If power is the suspect (as from cycling nearby equipment), you can try an additional line filter or you can move the device to another circuit, even if an extension cord is used.
- If facility grounding is the suspect, then you can improve the grounds with heavy wire (for low frequencies). If there is more than one module interconnected in the system, they you might try for a single-point ground, feeding all your equipment from a single-point ground and eliminating the rest of the grounds.
- If intrasystem grounding is the suspect, you can improve high frequency grounding by interconnecting with wide ground paths, such as with a strap of aluminum foil. Route your interconnect cables along this ground strap. This may also help if RFI is suspected.
- If RFI is the suspect, and it is forcible with a handheld radio, you can localize the problem by applying the radio in proximity to various elements of the system, such as the power and signal cables and to various parts of the enclosure. If cables (internal or external) are suspect, you can clamp on ferrites (use two or three to start with), use plug-in filters if available, or wrap aluminum foil around the suspect cables (terminating to the enclosure with conductive tape). If seams or openings are suspect, you can tape the seams shut (conductive mating is needed, so scrape the paint first). Ventilation holes can usually be closed for a trial with aluminum foil for a brief time without overheating. If you have a plastic box, you can enclose the entire box in aluminum foil for a trial.
- If ESD is suspected, the techniques are about the same as for RFI, with the addition that operator controls are a big suspect, especially if discharge can occur. Operator controls with exposed metals should be grounded firmly to the enclosure, with metal tape if necessary. The plate on which operator controls are placed also needs to be firmly grounded.

WHEN SYSTEM FIXES WILL NOT WORK

When you have inspected the installation, and have identified and corrected the problems at the facility level, then you may well be forced to go back to equipment modifications for a solution. Design issues have been discussed at some length in previous chapters, and we will give some brief guidelines for starting on redesign.

We have been in the EMI business for more years than we care to count, and we have seen some very subtle EMI problems. Nevertheless, most of the problems we encounter could have been avoided by following some basic steps. You will note that there is a significant overlap between the three cases. This is encouraging, because this means that a single fix will often work to solve two possible problems at one time. Additionally, this means that certain fixes can be applied even if you have not definitely pinned the problem down to a specific source.

RFI Problems

Here are some guidelines for preventing RF problems.

Use Multilayer Boards

The best type of circuit board for RFI problems is a multilayer board. Our experience shows that multilayer boards are at least 10 times better against RF than two-layer boards, a margin that is difficult to overcome with even a very good two-sided board layout. There are two reasons for this superiority–the ground plane underneath the traces virtually eliminates open loop areas (antennas), and the ground plane greatly reduces ground voltages on the board (ground bounce).

We often get the argument that multilayer boards are too expensive. We have also seen clients spend $50,000 in retesting and rework to avoid $10,000 in added manufacturing costs. Do not fall into this trap. Weigh the recurring circuit board costs against the nonrecurring costs of failure. Do not forget the "lost opportunity" cost if there is a schedule delay due to the two-layer board. Finally, if you have a very harsh environment or very sensitive circuits, you may even find it impossible to meet even modest requirements with a two-layer approach. Our advice: If you are using sensitive analog circuits, forget about two-layer boards.

Filter Critical Circuits

Once you have selected your circuit board, the next step is to filter critical circuits. These include analog circuits and also power supply circuits.

The sensitive analog circuit front ends are the most vulnerable for two reasons–first, they use low level signals, and second, they connect to the outside world. These circuits often have some other serious RFI constraints, such as high input impedance and leakage current limitations. High impedance inputs mean that series ferrites are not effective, unless shunt capacitors are used. Leakage current limits mean that you may not be able to "ground" capacitors to the case.

Analog output circuits should not be ignored, as they are also subject to RF jamming. They are much less sensitive than input circuits and they are usually much easier to filter.

You will also need multistage filtering if you have a broad frequency range to filter, since most filters can only cover about 2–3 decades of frequency. A maximum effort filter, with multiple stages for high and low frequencies and provisions to filter both CM and DM interference plus a capacitive termination for a cable shield, might be needed for very sensitive input circuits.

Power supply circuits must also include high frequency filtering. We are often asked why extra filtering is needed, since most power supplies already have a lot of filtering. Unfortunately, "ripple filters" are woefully ineffective at RF, so additional high frequency filtering is needed. Ferrites and small decoupling capacitors can work wonders here.

Keep Low Ground Impedance between Circuit Boards

Board interconnect is one of the weakest links in electronics. Provide for plenty of ground pins in the connector, or provide an adjacent ground strap. If you have fewer than one ground pin for every five signal lines, you are probably heading for trouble.

Use High Quality Cables and Connectors

Good quality cables and connectors are mandatory for RF shielding. Cables and connectors need to be considered as a team. You need a good quality braid shield, good quality connectors, and good connections between all the joints. We often use the "garden hose" analogy–leaks in the hose or leaks at the connections can cause problems. We recommend a high quality braid that mates to a metal connector circumferentially. This is particularly important at frequencies above about 10 MHz, where loose braids become very leaky.

Avoid "Pigtails" Like the Plague

Pigtails are a common "low frequency" technique for grounding cable shields introduces inductance in series with the shield that allows energy to couple to/from the shield and the inner wiring. This can be disastrous in the RF range, and is a leading cause of cable failures. Nevertheless, we continue to see this problem over and over again. A 1-cm pigtail has about 8 nanohenries of inductance, which represents an impedance of about 5 ohms at 100 MHz, which is way too high for a good "ground" connection. This can be avoided with a solid circumferential connection.

Use the Right Shielding Material

Radio frequency shielding is easy to achieve with the right materials. All it takes is a thin layer of highly conductive material, and even plated or painted surfaces provide high levels of shielding. Some cautions–carbon or graphite coatings do not work for RFI, and conductive plastics are often marginal. Magnetically permeable materials are not needed for good RF shielding, but are imperative for low frequency (audio) magnetic field shielding.

Since plastic enclosures are popular in medical devices, consider conductive paints (nickel or copper), vacuum plating, or electroless deposition. All three processes are widely used in the consumer industry with very good results. Do not overlook selective internal shields, if you want to protect selected circuits. The television industry has used this approach for years, putting sensitive "tuners" in small metal enclosed boxes.

Watch Out for Seams and Penetrations

Most RF shielding fails in relation to seams and penetrations. The problem is not where the shield is, but where it is not. As we have already seen, seams can act as "slot antennas" if they are greater than 1/20 of a wavelength long. At 100 MHz, this is about 15 cm, and this drops to 15 mm at 1 GHz. Incidentally, it is the longest dimension that is critical–a 6-inch seam with a 1 mm gap is as bad as a 6-inch diameter circle. The mating surfaces should provide good conductivity. Screw threads, hinges, and bearings make for poor conductivity.

Penetrations can also carry energy through the shield if they are not properly terminated. All cable shields must be grounded to the cabinet, and signal or power lines must be filtered to prevent high frequency currents from entering the shielded area. Cable shields and cabinet shields must work together to keep the unwanted RF out of the system.

ESD Problems

Ultimately, excessive ESD voltage/current must be blocked from sensitive components. Good design techniques center around making this happen, preferably in a cost-effective manner. If the adverse condition cannot be blocked by shielding, then the circuit itself must be protected.

Use Multilayer Boards

As with RFI, multilayer boards are clearly superior to double-sided boards, the principle reason being the stable ground layer. ESD must never be allowed onto a double-sided board: once ESD is on the board, it is practically impossible to cope with it.

If you must use double-sided boards, then you must provide a separate ground layer or shield in which to terminate your filter elements. Putting filters on a two-layer board to divert ESD is a futile effort.

Filter Critical Circuits

Once you have selected your circuit board, the next step is to filter (or transient protect) critical lines. For damage control every single line is at risk. If you cannot guarantee that the lines will be protected from direct discharge (notably when cabling up your system), then you must provide filtering or transient protection for each pin.

While ESD is a very fast and high current pulse, there is very little energy, so filter sizes can be quite small. You can set your cutoff frequency to several MHz, and still provide ample filtering for ESD. Do not go too small, though, as small capacitors will quickly charge up to unacceptably high voltages.

If you cannot filter high speed lines, then you might consider transient suppression using large geometry Zener diodes: MOVs and arc suppression devices are too slow to protect components from ESD.

Once cabled up, direct discharge into signal or ground pins is not the issue–the problem is discharge to nearby members, such as connectors and operator controls. In such a case you can reevaluate your filter requirements. Analog lines do not need ESD filters (note that they do need filters for RFI, however). Output lines are less susceptible to indirect ESD, and often do not need protection. Input digital lines are very vulnerable, and should always be filtered.

Terminating filters to the circuit ground is risky, although often a multilayer board is solid enough to absorb ESD currents without upsetting circuits on the board. If at all possible, terminate the filter to a shield boundary instead, as this is far and away the best answer.

While you are at it, apply bypass capacitors to all chips, especially microprocessors and other large-scale integrated devices. We find that battery-driven devices often lack such capacitors. Also, do not forget to bypass high impedance feedback circuits.

Keep Low Ground Impedance between Circuit Boards

Board interconnect is one of the weakest links in electronics. Provide for plenty of ground pins in the connector, or provide an adjacent ground strap. If you have fewer than one ground pin for every five signal lines, you are probably heading for trouble.

Use Metal Connector Backshells

Good quality connectors are mandatory for ESD protection. In particular, discharge directly to the connector shell can be devastating, as leaky connectors invite coupling of ESD energy into signal conductors. We recommend a metal connector circumferentially mated to the cable and to the mating connector. Mating of the shield to the backshell is imperative. Direct discharge to the shielded cable itself does not usually occur, so the threat to cables is generally due to indirect discharge, which is much less devastating. Mylar® film cable shields are adequate to cope with ESD, but it is difficult to mate to the shield to the connectors.

Avoid "Pigtails" Like the Plague

Pigtails, a common practice for grounding cable shields used in audio systems, are completely unacceptable for terminating ESD currents. Remember that a 2 cm pigtail (1 nH of inductance), provides an impedance of about 30 ohms at ESD frequencies. Thus, we see that a pigtail does little or nothing for ESD protection.

Similarly, avoid use of a drain wire on a Mylar® film shield. A widespread practice is to route the drain wire through the connector and terminate it inside, or worse yet, on the circuit board. Wrong answer! ESD currents pass through the drain wire coupling to immediately adjacent signal lines and corrupt the ground at the circuit board.

Use the Right Shielding Material

ESD shielding is easy to achieve with the right materials. A thin layer of highly conductive material will suffice. Plated plastic coatings (electroless or vacuum plating) provide ample shielding effectiveness for ESD. Conductive paints have a lower conductivity than that of plating, but are still usually adequate for ESD shielding. Carbon or graphite coatings are adequate for bleeding off charge to prevent ESD, but are inadequate to cope with ESD when it occurs. Conductive plastics are marginal, more because of the difficulty of achieving good mating contact than for the shielding properties themselves.

We have experienced some ESD problems with stainless steel, due to its low conductivity. Again, the problem lies not so much with the conductivity as with the inadequacy of the mating contact points. Magnetically permeable materials are not needed for ESD shielding, but are permissible.

Avoid Seams and Penetrations to the Shield

Shielding failure almost always occurs not because of the shield material, but because of seams and penetrations. Seams act like slot antennas, diverting ESD currents and provide some intense coupling paths in the immediate proximity of the seam. Using our 1/20 of a wavelength criteria, seams longer than about 2 cm are significant when it comes to coupling to inside electronics. While a longer seam does not automatically cause an ESD problem, connectors and cables (both internal and external) are especially vulnerable when in close proximity to seams (within about 2 inches).

Mating surfaces must provide good conductivity. Screw threads, hinges, and bearings are poor conductors. Specifically, using sheet metal screws to mate two surfaces is poor practice. Bearings, by the way, can be brought to premature failure if used as a ESD path: Each discharge pits a minute portion of the bearing—easily seen in a microscope.

Penetrations are conductors passing through the shield. Any ESD currents on the conductor will pass through the shield as if it were not there, going directly to the circuit board. Accordingly, all conductors must be terminated at the shield, either by shielding the cable or by terminating a filter at the shield boundary. Operator controls are especially vulnerable, so be sure to ground metal shafts to the enclosure or recess them so that discharge cannot occur.

Power Disturbance Problems

Ultimately, power disturbances must be blocked from sensitive components. Unlike ESD and RFI, power disturbance problems follow the power line, so the remedial action is much better defined. The path is from the power line through the power supply, and finally to the internal circuits. Your possibilities include intercepting the energy right at the power supply or at internal circuits. Starting from the inside out, the following can be done.

Protect Critical Circuits

In most applications there are a few circuits that are most vulnerable. The reset circuitry in your microprocessor system is very vulnerable, as is the feedback circuitry in your power supply. These have already been identified as being vulnerable to ESD and RFI, and the arguments are equally applicable to EFT. Protecting these circuits is necessary, filtering both the input lines and power feed to the chip.

We find the voltage supervisor is the biggest single problem with EFT. It is designed to detect failing power from your supply, but is unable to distinguish between a power loss and brief voltage sags or transients. Thus, the supervisor often does an unnecessary reset. We advise avoiding a power supervisor unless it is absolutely necessary. If your system requires one, then take extreme care in filtering this chip, along with all reset lines coming from off-board. Unfortunately, you still have little control over the minor sags that come from power supplies.

Filter Power Supply

Power supplies are typically not designed to protect against EFT. If you are designing your own supply, you can build in high frequency filtering. Purchased power supplies are not usually designed to comply with EFT requirements–you will probably have to add your own protection, in the form of filtering or transient protection.

Unfortunately, you usually do not have control over the feedback circuitry in your power supply, and transients will often result in a brief DC voltage sag. This condition must be intercepted at the front end of the supply.

Interference may well be CM, which is difficult to control where leakage currents are limited (as in patient-connected devices). Your only option is to use high impedance series filter elements, such as EMI type ferrites. Differential mode filtering is no particular problem.

Transient protection may be needed to control high amplitude transients. MOVs are commonly used, and they are effective for most power line disturbances. Large geometry Zener diodes (e.g., Transzorbs®) are also effective. Note that for full protection, line-to-ground and neutral-to-ground protection is needed, and this usually poses a conflict with safety requirements. Full transient protection is usually a facility issue.

SUMMARY

System problems can vary widely. Where possible, identify the potential problems before they occur. But, often the problem does not surface until the equipment is operational. But the issues can be generally placed into one or more of the following:

- *Identify threats.* Look for possible EMI threats at the site, including RFI, ESD, and power quality. Also look for sensitive equipment that may be adversely affected by your equipment, especially if you are a noise source.
- *Prepare the site.* Provide for dedicated clean power source to your equipment. If equipment involves a number of system elements, provide for single-point, low frequency grounds and for low impedance, high frequency grounds as appropriate. If shielding is needed, identify the shielding needs and either provide for a shielded enclosure or provide a shielded room.
- *Troubleshoot on-site.* If problems occur on-site, it may be necessary to isolate the problem on-site. Isolate your equipment, then filter, ground, or shield elements of the system as appropriate.
- *Fix at the factory.* If permanent solutions cannot be implemented on-site, return to the factory with a list of things to work on, as indicated by the nature of the problem on-site.

GLOSSARY

AC. Alternating current.

AM. Amplitude modulation.

ANSI. American National Standards Institute.

Aperture. Any opening in a conductive shield.

Band. A range of frequencies or wavelengths.

Bond. Electrical connection between conductive parts.

CE. Conducted Emissions. Term is used widely in military EMC testing.

CE Mark. Labeling affixed to equipment which is complaint to European Union requirements.

CISPR. International Special Committee on Radio Interference, a subcommittee of IEC.

CM. Common mode. Signal or interference which has the same polarity on both the signal or power line and the return line.

CMOS. Complimentary Metal-Oxide-Semiconductor.

Conducted. Energy or interference which is propagated by conductor.

Continuous Wave. Energy or interference which is continuous in nature.

CS. Conducted susceptibility. Term is used widely in military EMC testing.

CSA. Canadian Standards Association.

CW. Continuous Wave.

dB. Decibel. A ratio of power or voltages, expressed logarythmically.

dBμV/m. A unit of electric field strength, expressed relative to one microvolt.

DC. Direct Current.

DM. Differential Mode. Signal or interference which has the opposite polarity on the signal or power line. Sometimes called the normal mode.

EC. European Community. Now called the European Union.

ECL. Emitter Coupled Logic.

E-Field. Electric field, and which may be expressed in units of volts/meter or in dBμV/m or other variations, depending on magnitude.

EFT. Electrical Fast Transient. A high frequency burst of interference.

EFFT. Extremely Fast Fourier Transform. A fast method of determining frequency content of complex waves.

EMC. Electromagnetic Compatibility. Used to describe a condition wherein electrical and electronic equipment operate successfully together or in close proximity.

EMI. Electromagnetic Interference. Unwanted electrical energy which may impair function of electronic equipment.

Emissions. Electrical energy emanating from an electric or electronic source.

EN. European Norms. A prefix to European regulations, as in EN 601-1-2.

ESD. Electrostatic Discharge. Electrical discharge which follows electrostatic build-up.

EU. European Union.

Faraday Shield. Conductive element placed between two electrical elements to provide electric field isolation.

FAT-ID. Frequency, Amplitude, Time, Impedance, and Dimensions. An acronym for the key parameters in EMI.

FCC. Federal Communications Commission (U.S. government).

FDA. Food and Drug Administration (U.S. government).

Ferrite. Ferromagnetic element used to provide a lossy impedance element.

Finger Stock. Resilient serrated copper sheeting used to provide conductive closure between two conductive mating surfaces.

FM. Frequency Modulation.

Gasket. A conductive material used between two conductive surfaces to provide continuous conductive closure.

Ground. A return path for current.

Ground plane. A conductive sheet used for a ground.

HCMOS. High speed CMOS.

H-Field. Magnetic field expressed in Amperes/meter, Gauss, or dBpT.

IEC. International Electrotechnical Committee.

IEEE. Institute of Electrical and Electronics Engineers.

Immunity. Refers to immunity from EMI, usually established by test. This term is often used interchangeably with susceptibility.

ITE. Information Technology Equipment.

I/O. Input/output.

LISN. Line Impedance Stabilization Network. Used to establish a fixed impedance level when performing conducted interference tests.

Narrow band. Electrical activity which is constrained to a single frequency or a very narrow band of frequencies.

Radiated. Energy or interference which is propagated by radiation through space.

RE. Radiated Emissions. Term is used widely in military EMC testing.

RFI. Radio Frequency Interference. An older term for EMI, is now confined specifically to radio frequency interference.

RS. Radiated Susceptibility. Term is used widely in military EMC testing.

Seam. Junction between two metal elements in a shield or ground plane.

Shielding. Metallic closure used to reduce electric or magnetic field intensity.

Skin depth. Effective depth of current in a conductor. Varies inversely with increasing frequency, becoming very thin at high frequencies.

Surge. A surge of electrical energy, usually applied to a power line voltage.

Susceptibility. Susceptibility to EMI, usually as established by test. Term is often used interchangeably with immunity.

Termination. A term with refers to the method in which a transmission line or cable shield is connected at each end.

Transient. Interference which is not continuous, usually of a very brief duration, as in ESD or a lightning strike.

TTL. Transistor Transistor Logic.

V_{CC}. DC supply voltage for digital circuits.

Appendices

A

REFERENCES FOR EMC PUBLICATIONS

BOOKS ON DESIGN

Boxleitner, W. 1989. *Electrostatic discharge and electronic equipment.* New York: IEEE Press/Wiley-Interscience. Very good treatment of ESD design techniques, very simply written. Recommended as a first book.

Fluke, J. C. 1991. *Controlling conducted emissions by design.* New York: Van Nostrand Reinhold. Informative and fairly easy to read. Concentrates on conducted emissions. A companion to text by Mardiquian.

Gerke, D. D. and Kimmel, W. D. 1994. *EDN's designer's guide to electromagnetic compatibility.* Newton, MA: Cahners Publications. Originally published as a supplement to *EDN Magazine,* it is now available in reprints. This is a more general treatment of the subject matter contained in this book.

Interference Control Technologies 12 volume encyclopedia. 1988. A complete reference set. Expensive, worthwhile for the complete company library. Available from EEC Press, 6193 Finchingfield Rd, Box D, Gainesville, VA 22065, 703-349-4755. The company also publishes a number of other books.

Johnson, H. W. and Graham, M. 1993. *High-speed digital design.* Englewood Cliffs, NJ: PTR Prentice Hall. Concentrates on maintaining signal integrity on circuit boards and cables. Not specifically an EMI book, but it covers good circuit board design practices. Recommended for all circuit and logic designers.

Keiser, B. 1987. *Principles of electromagnetic compatibility.* Boston, MA: Artech House. Good general overview on EMI. Recommended as a first book. In the same genre as text by Ott.

Mardiguian, M., 1992. *Controlling Radiated Emissions by Design.* New York: Van Nostrand Reinhold. Informative and fairly easy to read. Concentrates on radiated emissions. A companion to text by Fluke.

Mardiguian, M, 1986. *Electrostatic discharge.* Gainesville, VA: EEC Press, 6193 Finchingfield Rd, Box D, Gainesville, VA 22065, 703-349-4755. Very good treatment on ESD phenomena and design techniques. Recommended as a first book.

Montrose, M. I. 1996. *Printed circuit board design techniques for EMC compliance.* IEEE Press. An in-depth treatment of printed circuit board layout and routing.

Morrison, R. 1986. *Grounding and shielding techniques in instrumentation.* New York: John Wiley & Sons. A complete treatment on grounding techniques, especially useful for instrumentation applications where low level analog signals are involved. A little bit hard to read.

Ott, H. W. 1988. *Noise reduction techniques in electronic systems.* New York: John Wiley & Sons. This book is the de facto standard reference book on EMI, quite readable. Recommended as a first book.

Paul, C. 1992. *Introduction to electromagnetic compatibility.* New York: John Wiley & Sons, Written as a college text book, it is very analytical. For the serious EMI student.

Smith, D. G. 1993. *High frequency measurements and noise in electronic circuits.* New York: Van Nostrand Reinhold. A good book on measuring interference to your equipment in the lab. Many quick measuring techniques and test anomalies are explained. This book has no near equivalent.

BOOKS ON ARCHITECTURE AND FACILITIES

Hemming, L. 1992. *Architectural electromagnetic shielding handbook.* New York: IEEE Press/Wiley Interscience. A complete guide for facility shielding.

Morrison, R. and Lewis, W. H. 1990. *Grounding and shielding in facilities.* New York: John Wiley & Sons. A guide for grounding, power and cable distribution and shielding in facilities.

Recommended practice for electric systems in health care facilities (IEEE White Book). 1986. ANSI/IEEE Std 602-1986. New York: IEEE Press/Wiley-Interscience. A complete guide for electric power systems in hospitals.

Recommended practice for emergency and standby power systems for industrial and commercial applications (IEEE Orange Book). 1987. ANSI/IEEE Std 446-1987. New York: IEEE Press/Wiley-Interscience. A complete guide for selecting and installing emergency and standby power equipment.

Recommended practice for power and grounding sensitive electronic equipment (IEEE Emerald Book). 1992. ANSI/IEEE Std 1100-1992. New York: IEEE Press/Wiley-Interscience. A complete guide for power and grounding practices for sensitive electronic instruments.

Recommended practice for the installation of electrical equipment to minimize electrical noise inputs to controllers for external sources. 1982. ANSI/IEEE Std 518-1982. New York: IEEE Press/Wiley-Interscience. A complete guide for installing industrial control systems.

PERIODICALS

Compliance Engineering. Compliance Engineering, 271 Great Dr, Acton, MA 01720; 508-264-4208. Bi-Monthly. General interest current EMC issues and other regulatory issues. Also an annual Reference Guide with same title. Free.

ENR (Electromagnetic News Report), Seven Mountains Scientific, PO Box 650, Boalsburg, PA 16827; 814-466-6559. Bi-monthly. A comprehensive newsletter on EMC design issues, regulatory affairs, product announcements and a calander of events. A modest subscription fee is required.

IEEE Transactions on Electromagnetic Compatibility, IEEE. Monthly. IEEE, PO Box 1331, Piscataway, NJ 08855; 800-678-IEEE. Highly analytical, generally written by researchers. Subscription is included with IEEE EMC Society membership. Not for the casual reader.

ITEM (Interference Technology Engineers' Master). R&B Enterprises, 20 Clipper Rd, West Conshohocken, PA 19428; 215-825-1960. Annual Reference Guide, plus quarterly update. Reference material for design and regulatory requirements. Free.

Kimmel Gerke Bullets. Kimmel Gerke Associates, Ltd. 1544 N Pascal, St. Paul, MN 55108; 612-330-3728. Authoritative quarterly newsletter by industry leaders. Free.

VENDOR REFERENCE MATERIAL

Electrostatic Discharge (ESD) Protection Test Handbook. KeyTek Instrument Corporation, 260 Fordham Road, Wilmington, MA 01887; 508-658-0880. A brief introduction to ESD phenomena and testing. Easy reading.

EMI Shielding Design Guide. Tecknit, 129 Dermody St, Cranford, NJ 07016, 201-272-5500. Reference on shielding and gasketing applications.

EMI Shielding Engineering Handbook. Chomerics, Division of Parker Hannifin,77 Dragon Court, Woburn, MA 01888, 800-225-1936. Reference on shielding and gasketing applications.

Fair-Rite linear ferrites catalog. Fair-Rite Products Corp, PO Box J, 1 Commercial Row, Wallkill, NY 12589; 914-895-2055. Reference on ferrite applications.

The Pulsed EMI Handbook. KeyTek Instrument Corporation, 260 Fordham Road, Wilmington, MA 01887; 508-658-0880. A brief introduction to power transients and testing. Easy reading.

Steward suppression ferrites, product and applications guide. D M Steward, PO Box 510, Chattanooga, TN 37401; 615-867-4100. Reference on ferrite applications.

Semiconductor houses have some good material on EMC. Look for the application notes, too. Motorola "MECL System Design Handbook" and National "FACT Advanced CMOS Logic Databook" are good, and Intel has some good application notes.

GOVERNMENT DOCUMENTS

MIL-STD-461D, 1993: *Electromagnetic emission and susceptibility requirements for the control of electromagnetic interference.* Military standard defining EMI requirements for military electronic equipment. Widely referenced in non-military electronics.

MIL-STD-462D, 1993: *Electromagnetic interference characteristics, measurement of.* Military standard detailing EMI test methods for the various requirements defined in MIL-STD-461D.

MIL-STD-419, 1982: *Grounding, bonding and shielding for electronic equipments and facilities.* Lots of grounding and shielding design information for assemblies, systems and installations. Very good reference.

MIL-HDBK-241B, 1983: *Design guide for electromagnetic interference (EMI) reduction in power supplies.* Very good for shielding and circuit design of power supplies.

AFSC DH1-4, 1984: *Air force systems command design handbook.* Good reference for EMI design. Hard to navigate through the document, however.

FIPS Pub 94, 1983: *Guideline on electrical power for ADP installations.* U.S. Dept of Commerce (NIST). Complete guide for facility power and grounding.

Most military documents may be ordered from Naval Publications & Form Center, Standardization Document Order Desk, 700 Robbins Ave, Bldg 4, Section D, Philadelphia, PA 19111-5094. Or fax order to 215-697-2978. They will not accept telephone orders, but you can call for information on 215-697-2667 or 215-697-2179. Most documents are free in single quantities. You need not be a government contractor to obtain most documents.

REGULATORY DOCUMENTS (IEC)

CISPR 11 (1990) Electromagnetic disturbance characteristics of ISM RF equipment

IEC 601-1-2 (1993) Medical electrical equipment

IEC 1000-4-2 (1995) Electrostatic discharge requirements

IEC 1000-4-3 (1995) Radiated EMC requirements for process control equipment

IEC 1000-4-4 (1995) Electrical fast transient/burst requirements

IEC 1000-4-5 (1995) Surge immunity requirments

IEC 1000-4-6 (1995) Immunity to conducted disturbances above 9 KHz

These documents are copyrighted. They are available from ANSI and other distribution channels.

B

PROPERTIES OF COMMON MATERIALS

DIELECTRIC CONSTANTS OF COMMON MATERIALS RELATIVE TO AIR

Material	ε_r	Material	ε_r
Air	1.0	Polyethylene/glass	4.5
Alumina	9.9	Polypropylene/cellular	1.5
Beryllia	6.8	Polypropylene/solid	2.2
Epoxy glass	4.8	Polystyrene	2.5
Epoxy resin	3.6	Polyurethane	7.0
Fused quartz	3.8	Polyvinylchloride (PVC)	3.5
Mylar® film	5.0	PTFE/glass	2.2
Nylon	3.0	RO-2800	2.8
Polimide/Kevlar®	3.6	Silicon Rubber	3.1
Polimide/quartz	3.3	Teflon	2.1
Polyethylene	2.3	Teflon/fiberglass	2.5
Polyethylene/cellular	1.7		

$\varepsilon_0 = 8.85 \times 10^{-12}$ farad/meter

CONDUCTIVITY/PERMEABILITY OF COMMON METALS

	Relative Conductivity	**Relative Permeability**
Material	σ_r	μ_r
Silver	1.05	1
Copper	1.00	1
Gold	0.7	1
Chromium	0.66	1
Aluminum (soft)	0.61	1
Aluminum (tempered)	0.4	1
Zinc	0.32	1
Beryllium	0.28	1
Cadmium	0.23	1
Brass	0.22	1
Nickel	0.20	100
Bronze	0.18	1
Platinum	0.18	1
Steel	0.17	1000
Tin	0.15	1
Lead	0.08	1
Monel	0.04	1
Conetic	0.03	25,000
Mumetal	0.03	20,000
Stainless Steel	0.02	500

σ (Cu) = 5.82×10^7 Siemens/meter

ELECTROCHEMICAL POTENTIALS OF COMMON METALS

Metal	Electrode Potential (volts)
Magnesium	-2.375
Beryllium	-1.700
Aluminum	-1.670
Zinc	-0.763
Chromium	-0.740
Iron	-0.441
Cadmium	-0.402
Nickel	-0.230
Tin	-0.141
Lead	-0.126
Copper	+0.346
Silver	+0.800
Platinum	+1.200
Gold	+1.420

SKIN DEPTHS OF COMMON SHIELDING MATERIALS

Frequency	Copper (inch)	Aluminum (inch)	Steel (inch)
60 Hz	0.335	0.429	0.034
100 Hz	0.260	0.333	0.026
1 kHz	0.082	0.105	0.008
10 kHz	0.026	0.033	0.003
100 kHz	0.008	0.011	0.0008
1 MHz	0.003	0.003	0.0003
10 MHz	0.0008	0.0003	0.0001
100 MHz	0.00026	0.0003	0.00008
1000 MHz	0.00008	0.0001	0.00004

$$\delta = \sqrt{\frac{2}{\pi f \sigma \mu}}$$

TRIBOELECTRIC SERIES

More Electropositive	
Air	Hard rubber
Human skin	Mylar® film
Asbestos	Epoxy glass
Glass	Nickel/copper
Mica	Brass/silver
Human hair	Gold/platinum
Nylon	Polystyrene foam
Wool	Acrylic
Fur	Polyester
Lead	Celluloid
Silk	Orlon® fiber
Aluminum	Polyurethane foam
Paper	Polyethylene
Cotton	Polyproplylene
Wood	Polyvinylchloride (PVC, vinyl)
Steel	Silicon
Sealing wax	Teflon
	More Electronegative

Note: Materials close together on the Triboelectric Series accumulate less electrostatic charge than those far apart. Nevertheless, charge will accumulate even with identical materials. At least one of the materials must be a dielectric, otherwise charge cannot readily accumulate.

PROPERTIES OF COPPER WIRE: SIZE AND RESISTANCE

AWG	Diameter (mils)	Ω/1000 ft (@ 20°C)	Max amps (@ 700 cm/amp)
8	128.5	0.6282	23.6
10	101.9	0.9989	14.9
12	80.81	1.588	9.33
14	64.08	2.525	5.87
16	50.82	4.016	3.69
18	40.30	6.385	2.32
20	31.96	10.15	1.46
22	25.35	16.14	0.918
24	20.10	25.67	0.577
26	15.94	40.81	0.363
28	12.64	64.90	0.228
30	10.03	103.2	0.144
32	7.950	164.1	0.090
34	6.305	260.9	0.057
36	5.000	414.8	0.036
38	3.965	659.6	0.022
40	3.145	1049	0.017

C

CONSTANTS AND CONVERSIONS

COMMON CONSTANTS AND CONVERSIONS

1 oersted = 1 gauss (μ_0 = 1 in cgs units)
$\mu_0 = 4\pi \times 10^{-7}$ henry/meter
$\varepsilon_0 = 8.85 \times 10^{-12}$ farad/meter
velocity of light (c) = 3×10^8 meter/sec
electronic charge (e) = 1.602×10^{-19} coulomb
1 tesla = 10^4 gauss
1 oerstad = 80 amp/meter
weber = volt-sec
farad = coul/volt
henry = weber/amp
oersted = maxwell/cm^2
tesla = weber/m

POWER AND VOLTAGE/CURRENT RATIOS CONVERTED TO dB

dB	Power Ratio	Voltage/ Current Ratio	dB	Power Ratio	Voltage/ Current Ratio
0	1.0	1.0	0	1.0	1.0
3	2.0	1.4	-3	0.50	0.71
6	4.0	2.0	-6	0.25	0.50
10	10.0	3.2	-10	0.10	0.32
20	10^2	10	-20	10^{-2}	0.10
30	10^3	32	-30	10^{-3}	0.03
40	10^4	10^2	-40	10^{-4}	10^{-2}
60	10^6	10^3	-60	10^{-6}	10^{-3}
80	10^8	10^4	-80	10^{-8}	10^{-4}
100	10^{10}	10^5	-100	10^{-10}	10^{-5}
120	10^{12}	10^6	-120	10^{-12}	10^{-6}

CONVERSION OF dBV, dBmV, AND dBμV

dBV	dBmV	dBμV
-120	-60	0
-100	-40	20
-80	-20	40
-60	0	60
-40	20	80
-20	40	100
0	60	120
20	80	140
40	100	160
60	120	180

CONVERSION OF dBμV TO dBmV AT 50 OHMS

dBmW	dBμV	dBmV
-100	7	-53
-80	27	-33
-60	47	-13
-40	67	7
-20	87	27
-10	97	37
0	107	47
3	110	50
6	113	53
10	117	57
20	127	67
40	147	87
60	167	107

CONVERSION OF GAUSS, TESLA, AND A/m

Gauss	Tesla	A/m
0.001	10^{-6}	0.07955
0.01	10^{-5}	0.7955
0.1	10^{-5}	7.955
1	10^{-4}	79.55
10	10^{-3}	795.5
100	10^{-2}	7955

VOLT/m TO mW/cm^2 FOR LINEAR AND dB SCALES

V/m	dBμV/m	mW/cm^2	dBmW/cm^2
1.00×10^{-06}	0	2.67×10^{-16}	-155.8
1.00×10^{-05}	20	2.67×10^{-14}	-135.8
1.00×10^{-04}	40	2.67×10^{-12}	-115.8
1.00×10^{-03}	60	2.67×10^{-10}	-95.8
1.00×10^{-02}	80	2.67×10^{-8}	-75.8
1.00×10^{-01}	100	2.67×10^{-6}	-55.8
$1.00 \times 10^{+00}$	120	2.67×10^{-4}	-35.8
$1.00 \times 10^{+01}$	140	2.67×10^{-2}	-15.8
$1.00 \times 10^{+02}$	160	2.67	-4.2
$1.00 \times 10^{+03}$	180	267	-24.2
2.00×10^{-6}	6	1.06×10^{-15}	-149.7
4.00×10^{-6}	12	4.24×10^{-15}	-143.7
6.00×10^{-6}	15	9.55×10^{-15}	-140.2
8.00×10^{-6}	18	1.70×10^{-14}	-137.7

FREQUENCY/WAVELENGTH CONVERSIONS

Frequency	λ	λ/2π
10 Hz	30,000 km	4,800 km
60 Hz	5,000 km	800 km
100 Hz	3,000 km	480 km
400 Hz	750 km	120 km
1 kHz	300 km	4.8 km
10 kHz	30 km	4.8 km
100 kHz	3 km	480 m
1 MHz	300 m	48 m
10 MHz	30 m	4.8 m
100 MHz	3 m	0.48 m
1 GHz	30 cm	4.8 cm
10 GHz	3 cm	0.48 cm

1/2π wavelength is boundary between near field and far field.

D

MISCELLANEOUS CONVERSIONS

RISE TIME/EQUIVALENT FREQUENCY

f(eq) = $1/\pi t_r$

t_r (ns)	f_{eq} (MHz)	Critical Length (in.)
1	318	3
3	106	9
10	32	30
30	11	90
100	3	300
300	1	900
1000	.3	3000

Above critical length, impedance termination is required and crosstalk becomes a potential problem.

RISE TIMES OF COMMON LOGIC FAMILIES

Family	ns	volt	volt	volt/ns
GaAs	0.1	0.8	0.1	8
ECL-100k	0.75	0.8	0.1	1.1
ECL-10k	2	0.8	0.1	0.4
AS	1.7	2.5	0.3	1.6
F	3	2.0	0.3	0.67
ALS	3	2.5	0.3	0.8
LS	3	2.2	0.3	0.7
AC	3	5.0	0.7	1.6
HC	6	5.0	0.7	0.8

Note: Worst of rise/fall time is assumed.

Index